ONE SIGNAL
PUBLISHERS

ATRIA

EVERYTHING IS PREDICTABLE

How Bayesian Statistics
Explain Our World

TOM CHIVERS

ONE SIGNAL
PUBLISHERS
———
ATRIA
New York London Toronto Sydney New Delhi

**ONE SIGNAL
PUBLISHERS**

ATRIA

An Imprint of Simon & Schuster, LLC
1230 Avenue of the Americas
New York, NY 10020

First One Signal Publishers/Atria Books hardcover edition May 2024

ONE SIGNAL PUBLISHERS / ATRIA BOOKS and colophon are
trademarks of Simon & Schuster, LLC

Simon & Schuster: Celebrating 100 Years of Publishing in 2024

For information about special discounts for bulk purchases,
please contact Simon & Schuster Special Sales at
1-866-506-1949 or business@simonandschuster.com.

The Simon & Schuster Speakers Bureau can bring authors to your live event. For
more information or to book an event, contact the Simon & Schuster Speakers Bureau at
1-866-248-3049 or visit our website at www.simonspeakers.com.

Interior design by Jill Putorti.

Manufactured in the United States of America

1 3 5 7 9 10 8 6 4 2

Library of Congress Cataloging-in-Publication Data has been applied for.

ISBN 978-1-6680-5260-0
ISBN 978-1-6680-5264-8 (ebook)

For Luis McGillycuddy, gone too soon,
and Mae Alison Davidson, recently arrived

Contents

A Theory of Not Quite Everything

The general rule in psychiatry is: if you think you've found a theory that explains everything, diagnose yourself with mania and check yourself into the hospital.[1]

—SCOTT ALEXANDER

Can you predict the future? Yes, of course you can.

You can predict with near certain accuracy that in the next few seconds, you'll take a breath, and let it out again. Your heart will beat, somewhere between one and three times a second. Tomorrow morning, the sun will come up, at a particular time, which depends on your latitude and the time of year, but which nonetheless you can find out with great accuracy. All of these events you can predict with confidence.

You can also predict that the train will arrive at a certain time, or that your friend will arrive on time at the restaurant at which you've arranged to meet them. Though, depending on the rail company, or your friend, you might be less confident in that.

And you can predict that the world's population will continue to grow until around the middle of the century, and then start to fall again. You can predict that global average surface temperatures in the year 2030 will be higher than they were in the year 1930.

The future isn't opaque. You can see into it. Some parts are more predictable than others—the Newtonian dance of the planets we can predict out for thousands of years; the Lorenzian chaos of the weather, really only a few days. But you can peer through the murk, after a fashion.

That's not what people normally mean when they say, "I can predict the future." They are referring to something mystical, some psychic or magical vision. We probably can't do that. (You'll read about a scientist in this book who thinks we can, and you'll also read about why he's almost certainly wrong.) But we don't need to. All that we do, all the time, is predict the future. We couldn't function if we couldn't. We make very basic predictions, like "the air will continue to be breathable," implicitly, with every breath we take. We make more complex predictions, like "The corner shop will have granola when I get there," each time we make a decision. We're not basing them on mystical visions, but on information we have gathered in the past.

The thing with all these predictions is that they are *uncertain*. The universe may or may not be deterministic; perhaps if we had perfect, godlike knowledge of the position, movement, and qualities of every particle in the universe, we could perfectly predict everything, the fall of every sparrow. But we don't. Instead, we have partial information. We can see bits of the universe, imperfectly, through our imperfect senses. We have best guesses for the way those bits move—we know the human-shaped bits tend to seek food and company; we know the rock-shaped bits tend to sit still. We can make messy, imperfect predictions with that information.

Life isn't chess, a game of perfect information, one that can in theory be "solved." It's poker, a game where you're trying to make the best decisions using the limited information you have.

This book is about the equation that lets you do that.

"Someone told me," said Stephen Hawking after the publication of *A Brief History of Time*, "that each equation I included in the book would halve the sales."[2] This book is about an equation, so it will be difficult to avoid including at least one.[*]

That equation is Bayes' theorem, or Bayes' rule. As equations go, it is simple. It looks like this:

$$P(A|B) = \frac{P(B|A) \cdot P(A)}{P(B)}$$

My dirty little secret is that I hate reading equations. I can do it, sort of. But it's a slog. That's embarrassing because I have now written three books that are either entirely or partly about math. But my brain grinds to a halt at the sight of a Σ symbol. And I suspect that quite a lot of readers feel the same, which is probably why Hawking was warned about including them in his book.

But equations aren't secret codes or arcane magic. Each little symbol (I have to remind myself) denotes a simple action. It's just a sort of shorthand.

In this case, Bayes' theorem is about probability: about how likely

---

[*] It's nice to think, though, that had I managed it, I might have sold *four* copies

something is, given the evidence we have. Specifically, it's about a particular form of *conditional* probability. The vertical line | is shorthand for "in the event that" or "conditional on." So P(A|B) is the probability of an event A happening, given that event B has happened.

Here's a simple example of conditional probability: say you wanted to know the probability of drawing a heart from a deck of cards. You know there are thirteen hearts in a standard fifty-two-card deck, so your probability—P($\heartsuit$), if you like—is ¹³⁄₅₂, or ¼. Or, in probability notation, p = 0.25. But then you draw a card, and it's a club. What's your probability now? Well, there are still thirteen hearts in the deck, but only fifty-one cards in total. So your probability is now ¹³⁄₅₁, or p ≈ 0.255. (The wavy equals sign means "approximately equal to.") That's the probability of drawing a heart *given that* you've previously drawn a club, P($\heartsuit$|$\clubsuit$).

Or: What's the probability that it will rain on a given day in London? Probably about 0.4: there are around 150 rainy days a year in London. But you look out the window and you see that the clouds are dark and heavy. What's the probability now? I don't know exactly, but higher: the probability of rain *given that it's cloudy* is higher.

Bayes' theorem is the same idea, but taken a bit further. In natural language it means: the probability of event A, given event B, equals the probability of B given A, times the probability of A on its own, divided by the probability of B on its own.

Imagine that you have a disease that is spreading through your society. That shouldn't be too hard to imagine, given recent history.

You want to know whether you have the disease, so you take a test. On the instructions that come with the test, there's a little note: "This test is 99 percent sensitive and 99 percent specific." What that means is that if you have the disease, there's a 99 percent chance that the test will, correctly, tell you that you have the disease; if you don't have the disease, there's a 99 percent chance that it will tell you, correctly, that you don't have the disease. Another way of saying this is that the test has a "false negative rate" of 1 percent, and a "false positive rate" also of 1 percent.

So you take the test, and you get a positive result: two lines show up. What does that mean? You might, reasonably, assume that it means it's 99 percent likely that you have the disease.

But it doesn't. And the reason it doesn't is Bayes' theorem.

Bayes' theorem is strange. It is a simple equation, which you can write on a line, and which is composed only of mathematical operations that most eight-year-olds could carry out—multiplication and division. It was first worked out by an eighteenth-century gentleman hobbyist, a part-timer whose day job was being a Nonconformist minister in Tunbridge Wells, England. But it has profound implications—it's why a cancer test can be 99 percent accurate even though 99 percent of the people it says have cancer don't; it tells us why DNA forensics might only have a 1 in 20 million chance of wrongly matching an innocent suspect, but still be more likely than not to send the wrong person down. It explains why scientific results can be "statistically significant" and yet still very probably wrong.

Bayes' theorem also reveals fascinating philosophical divides. Is "probability" a real thing? When we say that there's a one in six chance that we'll roll a one, what do we mean? Is that some fact about the universe, or just a statement about our beliefs in the world? And can one-off events have probabilities? If I say there's a 90 percent chance that Man City soccer club will win the league in 2025, what does that mean?

When we make decisions about things that are uncertain—which we do all the time—the extent to which we are doing that well is described by Bayes' theorem. Any decision-making process, anything that, however imperfectly, tries to manipulate the world in order to achieve some goal, whether that's a bacterium seeking higher glucose concentrations, genes trying to pass copies of themselves through generations, or governments trying to achieve economic growth: if it's doing a good job, it's being Bayesian.

Artificial intelligence is essentially applied Bayes. It is, at its most basic level, trying to predict things. A simple image classifier that looks at pictures and says they're of a cat or a dog is just "predicting" what a human would say, based on its training data and the information in the picture. DALL-E 2, GPT-4, Midjourney, and all the other extraordinary AIs that are wowing people as I write, the things that can hold conversations with you or create astonishing images from simple text prompts, are just predicting what human writers and artists would make from a prompt, based on their training data. And the way they do it is Bayesian.

Our brains are Bayesian. That's why we are vulnerable to optical illusions, why psychedelic drugs make us hallucinate, and how our minds and consciousnesses work at all.

And Bayes' theorem can help us understand why conspiracy theories are so hard to shift, and why two people can look at the same evidence and have it tell them entirely different things. Why is it that skeptics look at the scientific evidence that convinces me that vaccines are safe and effective, and be unmoved by it? It's because, as dictated by Bayes' theorem, your response to new information is influenced by the beliefs you already hold. It's not that vaccine skeptics or conspiracy theorists are strange aliens whose brains work differently: it's that they are behaving entirely rationally, given their existing beliefs. And Bayes' theorem explains how that works.

It is a theory of not-quite-everything, perhaps. *Nearly* everything. Once you start looking at the world through a Bayesian lens, you do start seeing Bayes' theorem everywhere. My intention is to make you see it everywhere too.

The usual way to explain Bayes' theorem is with medical testing. Here's a realistic example with plausible numbers: you are going for breast cancer screening. You know that if a woman has cancer, the mammogram will correctly identify it 80 percent of the time (it's 80 percent sensitive) and miss it the other 20 percent. If she *doesn't* have cancer, it will correctly give the all clear 90 percent of the time (it's 90 percent specific), but give a false positive 10 percent of the time.

You get the test. It comes back positive. Does that mean there's a 90 percent chance you've got breast cancer? No. With the information I've given you, you simply don't know enough to say what your chances are.

What you need to know is how likely you thought it was that you had breast cancer *before* you took the test. One simple way of guessing that is finding out what percentage of women your age have breast cancer at any given time. Let's say it's 1 percent.

To keep things concrete, let's imagine 100,000 women get tested. Of those 100,000, 1 percent—1,000—actually have cancer. Of those

1,000, the test will correctly diagnose 800 of them—80 percent—but falsely give the all clear to 200. Of the 99,000 who don't have cancer, it will correctly give the all clear to 89,100, but falsely diagnose cancer in 9,900. Or, in table form:

	HAVE CANCER (1,000)	DON'T HAVE CANCER (99,000)
TEST POSITIVE	80 PER CENT (TRUE POSITIVES) 800	10 PER CENT (FALSE POSITIVES) 9,900
TEST NEGATIVE	20 PER CENT (FALSE NEGATIVES) 200	90 PER CENT (TRUE NEGATIVES) 89,100

So now you can tell. You walk into an oncologist's office and get a positive mammogram. Of the 10,700 women who got a positive result, 800 actually had cancer. So your chance of really having cancer, if you get a positive result in this case, is 800/10,700 ≈ 0.07, or about 7 percent.

But this is entirely dependent on how likely you were to have cancer in the first place. If you were testing a higher-risk population—say, older women with a family history of cancer—it might be that 10 percent of the women you're testing have cancer. Then the math changes dramatically:

	HAVE CANCER (10,000)	DON'T HAVE CANCER (90,000)
TEST POSITIVE	80 PER CENT (TRUE POSITIVES) 8,000	10 PER CENT (FALSE POSITIVES) 9,000
TEST NEGATIVE	20 PER CENT (FALSE NEGATIVES) 2,000	90 PER CENT (TRUE NEGATIVES) 81,000

Now, instead of 800 true positives, you have 8,000. And your number of false positives has gone down to 9,000. So the chance you've got cancer is 8,000 divided by 17,000, or about 47 percent, a much more worrying prospect. The test hasn't changed; all that's changed is the prior probability.

What Bayes' theorem does is tell you how much you should change your belief. But in order to do that, you have to have a belief in the first place.

To go back to the equation (it won't halve my sales again, I've already used it):

$$P(A|B) = \frac{P(B|A) \cdot P(A)}{P(B)}$$

What it gives you, once you run the numbers, is P(A|B): the probability of A, the event, given B, the evidence. So the probability of having the disease, given a positive test result. That's all you're really interested in: I've got the result, so how likely is it that I have the disease?

But what the "80 percent sensitive" statistic gives you is *the exact opposite*. It's P(B|A), the probability of B, given A. It answers this question for you: How likely am I to see this result, given that I have breast cancer?

It might sound unimportant, but it's the difference between "There's only a 1 in 8 billion chance that a given human is the pope" and "There's only a 1 in 8 billion chance that the pope is human."[3]

In order to work out the thing we *really* want to know, we need more information. In the example of the cancer test, we need to know how common breast cancer is in the population being tested. In medical terms, that's the prevalence or the background rate, but in Bayes' theorem in general, it's known as your *prior probability*, or "prior."

In medical testing, your prior is often relatively easy to work out, or at least straightforward to define. If you're trying to work out someone's risk of Huntington's disease, you can look up diagnoses recorded in general practice records[4] and estimate that about 12.3 people per 100,000 have it.

For other situations, it's much more difficult. If you want to know how likely it is that Russia will invade Ukraine, what's your prior probability? How often Russia has invaded Ukraine per year? How often one country invades another? How often one country invades another when they have just sent a whole load of tanks to that country's border?

Take another example. How likely is it that this scientific hypothesis of mine is true, given that I've just done an experiment and seen some particular data? Let's say that, if my hypothesis was false, I'd only expect to see data like this one time in every twenty. Does that mean I can say that the hypothesis is probably true? No—it depends on how probable my hypothesis was before I began my experiment, my prior probability. But how on earth do I work that out?

And another. How likely is it that this person is guilty, given some forensic evidence? If I've got DNA evidence that would only show up by chance one time in a million, does that mean there's only a one-in-a-million chance that I've got the wrong suspect? No: it depends how likely it was that you had the right suspect in the first place. Again: How do you even start to put numbers on these things?

We'll get into all that. (There are people who do it for a living.) But the important thing is that you have to start with a prior probability, and use Bayes' theorem. If you don't, you end up in some strange places.

The first place most people come across Bayes' theorem is in medicine, so let's start there.

I've been mildly obsessed with Bayes' theorem for years. I first read about it in Ben Goldacre's "Bad Science" column for the *Guardian* in the early 2000s. Since then, I've steadily become more fascinated. I've written three books, including this one, and Bayes makes an appearance in all of them. There's something wonderful about how counterintuitive the theorem is. What do you *mean*, a test being 99 percent accurate isn't the same as a 99 percent chance that it's right? What mad language are you talking? If you follow the really-not-that-difficult reasoning, it becomes clear, but—for me, at least—it never quite loses its uncanny, otherworldly feeling.

But over the last three years, since early 2020, when COVID-19 started marching across the world, it became much more salient. Way back in April 2020, when we were still deep in the first lockdown, people, including former UK prime minister Tony Blair, were calling for "immunity passports," antibody tests that could tell if someone had had COVID or not. If they had, those people should be allowed out and about. (This was back before we realized you could get multiple infections pretty easily.)

At the time, antibody tests were just coming out. One that had just been issued emergency approval in the US reported roughly 95 percent sensitivity and specificity. [5]

Which sounds pretty good. But in April 2020, probably about 3 percent of British people had had the virus. That's your prior probability. If you tested a million people with this test, you'd expect about 30,000 to actually have had COVID. Your test would correctly identify about 28,500 of them. But of the 970,000 people who *hadn't* had COVID, it would incorrectly say that 48,500 of them *had* had COVID.

So of the 77,000 positive results you'd probably get, little more than a third would really have had the disease. (That's your *posterior probability*.) If you had tested all 65 million Britons, and issued "immunity passports" to everyone who got a positive result, it would have meant telling about 3 million people that they were safe to go and hug their immuno-compromised grannies when they very much weren't. You just couldn't have made any sense of this without some sort of grasp of Bayes.

In Britain, there was another Bayes-related controversy when a few members of the "lockdown-skeptical" commentariat became dimly aware of it. The former government minister John Redwood was probably the most famous: he demanded that "government advisers today need to tell us how they are going to stop false test results distorting the figures."[6]

What had happened was that one of them had misinterpreted an interview with Professor Sir David Spiegelhalter, a cheerful statistician who spent a lot of time on national TV and radio during the pandemic patiently explaining testing accuracy or vaccine efficacy. They worked out that just because a test has a 1 percent false positive rate, it doesn't mean that only 1 percent of positives are false. That was back between the first and second waves, when we were all doing polymerase chain reaction (PCR) tests every time we thought we had the sniffles. At the time, the prevalence of COVID in the British population was pretty low—lockdowns reduce infections!—but seemed to be creeping back up.

But the COVID contrarians thought that the apparent increase was an illusion that could be explained away with Bayes' theorem. About 0.1 percent of people had COVID at the time. If you tested people at random, and your test correctly identified people who didn't have COVID 99 percent of the time, and people who did have COVID 90 percent of the time, more than 90 percent of your positives would be false.*

This is all completely true. But they hadn't pushed the Bayesian reasoning far *enough*. First: Is the prior probability really 0.1 percent? Sure, if you're testing the population completely at random. But we weren't: we were testing people *who had symptoms* or who had come into contact with a confirmed case. Those people would be much more likely to have the virus. How much more likely? We don't know, but even if only 1 percent of them genuinely had COVID, the percentage of your total positives that are false drops to 50. If 10 percent of them do, about 90 percent of your positive test results will be real.

And, of course, we're assuming that the false positive rate really is 1 percent. That seems amazingly unlikely. At one point in summer 2020, when COVID had died down a bit, the *total* percentage of tests coming back positive, whether false or true, was 0.05 percent, so the false positive rate can't reasonably have been higher than that. If we use that, then with a COVID prevalence of 0.1 percent, your false positives drop to about 35 percent. If we assume that the prevalence in the testing population was higher, for the reasons outlined above, then it would be lower still.

---

* You test 1 million people. Of them, 1,000 actually have COVID. Your test identifies 900 of them. Of the remaining 999,000, it incorrectly diagnoses 9,990 as having COVID. 900 + 9,990 = 10,890. 900 is about 9 percent of 10,890.

But it's not just COVID. You can't make sense of pretty much any form of medical testing without invoking Bayes.

The NHS in England offers three kinds of routine cancer screening: breast, cervical, and colon. Prostate screening is available for men over fifty if they ask for it, but it's not routinely offered. Why not? Cancer screening just sounds like a good thing. We all know that early detection improves outcomes. Why wouldn't you want to do a test that tells you if you've got cancer or not?

The answer, as with everything in this book, can be found in Bayes' theorem.

Prostate cancer screening is carried out with something called a prostate-specific antigen (PSA) test. It's pretty simple. You get a blood test, and if the levels of PSA in your blood are above a certain level—usually three or four nanograms per milliliter—then you're sent for further testing, such as a scan or a biopsy. High PSA can be a sign of prostate cancer, although it can also be a sign of infection, inflammation, or just age.

PSA screening is not as accurate as the tests we've been talking about so far. According to the National Institute for Health and Care Excellence (NICE), the UK's medical advisory body,[7] if you were to screen for PSA with a cutoff of three nanograms per milliliter, then it would correctly identify about 32 percent of patients with cancer (sensitivity) and about 85 percent of cancer-free patients (specificity).

About 2 percent of men in their fifties have prostate cancer.[8] If you tested a million patients again, about 20,000 of them would actually have cancer. You'd correctly identify about 6,400 of them. And of the remaining 980,000, you'd tell about 147,000 that they needed a follow-up check. If you got a positive result on this test, as a man in your fifties, there'd only be about a 4 percent chance you actually had cancer.

Is a 4 percent chance worth knowing about? Maybe. But bear in mind you'd need extra tests, some of which are invasive, unpleasant, and somewhat risky. Plus, of course, the NHS would have to pay for tens of thousands of MRI scans and biopsies, at a cost of some millions of pounds, money that could have been spent on statins or kidney transplants or nurses' wages. And the thing about prostate cancer is that, in many cases, it's so slow growing that men don't know they have it; very often, men are found to have prostate cancer in postmortem examination, having died of something else entirely.

This also raises another important point. The 32 percent sensitivity/ 85 percent specificity figures come from using a three-nanograms-per-milliliter cutoff. But you could bump it up to four nanograms. What happens then?

Well, you get more specificity. The percentage of cancer-free patients that a test correctly identifies as cancer-free goes up from 85 to 91 percent. But that comes at a cost in sensitivity. The percentage of men with cancer whom it correctly identifies goes down from 32 percent to 21 percent. If you tested your million men again, now you'd get fewer false positives—down to 88,200—but fewer true positives as well: just 4,200 out of the 20,000. In that situation, if you got a positive result, you'd still only have about a 4.5 percent chance of actually having cancer.

You can't get around this. You can move the threshold up—have the cutoff at five nanograms per milliliter, say—and you can reduce the number of false positives, but only at the cost of increasing the number of false negatives. Or you can move the threshold down, and decrease the false negatives, but only at the cost of more false positives. It's an unavoidable trade-off, cast in stone. The only way around it is to use a different, better test. (This is analogous to the problem of "statistical significance" in science, which we'll come back to later.)

In breast cancer and colon cancer, the screening is rather more accurate. But even there, it's highly dependent on the prevalence of the disease in the population. One major study[9] found that 60 percent of women who have annual mammograms for ten years get at least one false positive result, leading to referrals for extra investigations such as biopsies and causing "anxiety, distress, and breast cancer–specific worry." Is that worth it? It entirely depends on the background rate of the disease in the population: your prior probability. Breast cancer is rare among the young. If you test women under forty, even quite sensitive and specific tests end up with very high numbers of false positives. Among older women, it becomes more valuable, and NICE says that it is cost-effective in women over fifty.[10] But you can't make decisions about it without Bayes.

Would-be parents would do well to read up about Bayes as well. There's a kind of antenatal screening known as non-invasive prenatal testing, NIPT, in which a blood sample is taken from a pregnant woman and tested for various chromosomal conditions in the fetus. In the UK, the NHS offers it to women in higher-risk categories. But it's also available, for about £500 ($600), through private clinics.

The test is sold as being 99 percent accurate. But once again, the accuracy of the test on its own doesn't tell you anything about how likely it is that your result is correct. The conditions it tests for—Down's syndrome, Patau's syndrome, and Edwards syndrome—are all rare. They're also very serious. A child with Down's can lead a long and happy life, but will often require lifelong care, while those with Patau's and Edwards usually die in their first months or years of life. It obviously matters a great deal to parents whether their test results are accurate or not.

A review of the evidence found[11] that doing NIPT tests on the general population, rather than limiting it to high-risk pregnancies, often gave false positives. The "positive predictive value"—that is, the percentage chance that a given positive was a true positive—for Down's syndrome was 82 percent, for Patau's syndrome 49 percent, and for Edwards syndrome just 37 percent.

If you limited your scope to just the high-risk categories, those numbers rose significantly—for Edwards, the positive predictive value jumps to 84 percent. That is, if you run the test on mothers-to-be at random, then nearly two out of every three positive results you get will be false. But if you limit it to just those at higher risk, fewer than one in six will be.

Again, this is pure Bayes. Your new data on its own cannot tell you the whole story. You need to know your prior probability. It's not a hypothetical or academic problem. If you're expecting a baby, and you do one of these tests and get a positive result, Bayes' theorem is central to your decision about what to do next. And, as we'll learn later, you can't necessarily expect your doctors to be able to help you. Doctors, just like the rest of us, tend to assume that a 99 percent accurate test is right 99 percent of the time.

It's not just medicine. In law, there's a thing called the prosecutor's fallacy, which is quite literally just *not thinking like a Bayesian*. Imagine you do a DNA test on a crime scene. You find a sample on the handle of the murder weapon that matches the DNA of someone in your database. The DNA match is quite precise—you'd only expect to see a match that close about one time in every 3 million.

So does that mean that there's only a one-in-3-million chance that your suspect is innocent? By now, hopefully, you'll have realized that's not the case.

What you need to know is your prior probability. Is there any particular reason to think this person is the right one, or is your database just a random selection of people from the British population? If so, then your prior probability that the person you're accusing is the criminal is one in about 65 million: there are 65 million Britons and only one person who committed this particular crime. If you DNA-tested every Briton, you'd get about twenty DNA matches, just by chance, plus the perpetrator. So the probability that you've got the right suspect is about 5 percent, give or take.

But if you had narrowed it down to just ten suspects beforehand—say that you're Hercule Poirot and you know it's one of ten people trapped in a country mansion by a snowstorm—then it's very different. Your prior probability is 10 percent. If one of *those* ten people match the DNA, then your probability of a false positive is about one in three hundred thousand.*

Once again, this is not some pettifogging point. Real court cases have turned on these details. In 1990, a man called Andrew Deen was

---

*Obviously, that doesn't necessarily mean it's a 1-in-300,000 chance they're innocent—their DNA may have got on the weapon in some other way than them being the murderer.

convicted of rape partly on the basis of DNA evidence. An expert witness told the court that the chance that the DNA came from someone else was just one in 3 million. But Deen's conviction was overturned (although he was convicted in his retrial) because, as a statistician explained,[12] the two questions "How likely is it that a person's DNA would match the sample, if they are innocent?" and "How likely is it that someone is innocent, given that their DNA matches the sample?" are not the same, just as "How likely is it that a given human is the pope?" is not the same as "How likely is it that the pope is a human?"

Sometimes, the errors go the other way. During the trial of O. J. Simpson, the former American football star, for the murder of his wife, Nicole Brown Simpson, the prosecution showed that Simpson had been physically abusive. The defense argued that "an infinitesimal percentage—certainly fewer than 1 in 2,500—of men who slap or beat their wives go on to murder them"[13] in a given year.

But that was making the opposite mistake to the prosecutor's fallacy. The annual probability that a man who beats his wife will murder her might be "only" one in twenty-five hundred. But that's not what we're asking. We're asking if a man beats his wife, and *given that the wife has been murdered*, what's the probability it was by her husband?

Gerd Gigerenzer, a German psychologist and scholar of risk, pointed out that if that one in twenty-five hundred figure is right, then for every one hundred thousand women who suffer domestic abuse, about forty will be murdered.[14] The base rate for murders among American women is about five in one hundred thousand.

So the prior probability of an American woman who is a victim of domestic abuse being murdered by her husband is about one in twenty-five hundred per year. But we need to update that probability with new information—we now know that the woman in question *was* murdered.

We can now do the Bayesian maths. If we take 100,000 domestic abuse victims, then, presumably, in a given year, 99,955 are not murdered. But of the remaining 45, 40 are murdered by their husbands. The defense had made the inverse of the prosecutor's fallacy: they had used just the prior probability, and ignored the new information coming in.

Bayes' theorem, while it helps us understand these errors of reasoning, can tell us more profound things too. The word "inverse" in the last paragraph is key. Often, statistics and probability will tell you how likely it is that you'll see some result by chance. If my dice are fair, I'll see three sixes at the same time 1 time in every 216. If I was never at the crime scene, my DNA should match the sample 1 time in every 3 million.

Often, though, that's not what we want to know. If we're worried that the person we're playing craps with is a cheat, we might want to know "If he rolls three sixes, what are the chances that his dice are fair?" If someone's DNA matches the sample at the crime scene, we might want to know what the chances are that it's a fluke. And that is the exact opposite question.

For quite a long time, the history of probability was about asking the first question. But after the Reverend Thomas Bayes—about whom much more later—started asking the second one, in the eighteenth century, it became known as *inverse probability*.

As you'll see over the course of this book, it's strangely controversial. Bayes' theorem has devotees and enemies, far more than any comparable one-line equation. You don't get people yelling at one another online over the formula for the surface of a sphere, or over Euler's identity equation.

But I think that's because it affects everything. How likely is a scientific hypothesis to be true, given the result of some study? Well, I can tell you the probability that you'd see the results we've seen if it *weren't* true, but that's not the same thing. To estimate how likely it is—and a growing number of scientists argue that that's exactly what we want statistics to be doing—we need Bayes, and we need prior probabilities.

More than that, *all* decision-making under uncertainty is Bayesian—or to put it more accurately, Bayes' theorem represents ideal decision-making, and the extent to which an agent is obeying Bayes is the extent to which it's making good decisions. Logic itself, all that stuff you may remember about "All men are mortal; Socrates is a man; ergo Socrates is mortal" is just a special case of Bayesian reasoning where you're only allowed to use probabilities of one and zero.

We appear to be Bayesian machines. That's true at a fairly high level: humans are rubbish at working out Bayes' theorem formally, but the decisions we make in everyday life are pretty comparable to those that an ideal Bayesian reasoner would make. Which, unfortunately, doesn't mean we all end up agreeing—if my prior beliefs are very different from yours, then the same evidence can lead us to entirely different conclusions. Which is how we can end up with profound, but sincere, disagreements on apparently well-evidenced questions about the climate, or vaccines, or any number of other questions.

And we're Bayesian at a deeper level too. Our brains, our perception, seem to work by predicting the world—prior probabilities—and updating those predictions with information from our senses: new data. Our conscious experience of the world can be best described as our priors. I predict, therefore I am.

CHAPTER ONE

From *The Book of Common Prayer* to the Full Monte Carlo

BAYES THE MAN

Near Old Street Tube station, in Shoreditch in East London, there is a graveyard known as Bunhill Fields.

Quite a few well-known people are buried in Bunhill. William Blake is perhaps the most famous, or Daniel Defoe, author of *Robinson Crusoe* and *A Journal of the Plague Year*. John Bunyan, author of *The Pilgrim's Progress*, is buried there too.

But for the sort of person who would, as I have on several occasions, be walking from the Tube to the nearby Royal Statistical Society, Bunhill is best known for being the final resting place of the Reverend Thomas Bayes.

Bayes was an eighteenth-century Presbyterian minister and a hobbyist mathematician. In his lifetime, he wrote a book about theology and another about Newton's calculus. But what he is remembered for is his short work, "An Essay towards Solving a Problem in the Doctrine of Chances."[1] It was published posthumously, in the journal *Philosophical Transactions*, after his friend Richard Price found and edited some unfinished notes Bayes left behind.

This book is about the deceptively simple idea that Bayes came up with, his theorem. It is, without exaggeration, perhaps the most important single equation in history. But very little is known about the man himself. The fact that we can only say he was *probably* born in 1701 gives you an idea of how hazy our knowledge is.

David Bellhouse, an emeritus professor of statistics at the University of Waterloo in Canada, wrote a biography of Bayes for the journal *Statistical Science*[2] in 2004. The problem, he says, was that Bayes was a Nonconformist: a member of a church that dissented from the teachings of the Church of England.

To explain why that's a problem, we have to go back a couple of centuries. Fans of the Hilary Mantel novel *Wolf Hall* will remember that Henry VIII took England out of the Catholic Church in 1533, in order to marry Anne Boleyn. He died in 1547, several wives later, and after his death Archbishop Cranmer introduced *The Book of Common Prayer* in 1549, making it obligatory for all English churches to use it in their services.[3]

Henry's daughter Mary disagreed with that decision and abandoned it in 1553, having Cranmer burned at the stake for heresy to drive the point home. Then Elizabeth I reinstated the decision a few years later, and everyone carried on using the *Book* for nearly a century, until the English Civil War.

During the period of the Commonwealth, from the execution of

Charles I in 1649 until the restoration of the monarchy in 1662, the restrictions on forms of worship were relaxed; but in 1662, Parliament passed an Act of Uniformity, requiring that the *Book* be used in all services in England once more.

By now, some clergymen were used to the freedom they had enjoyed under Oliver Cromwell's Commonwealth. About two thousand of them refused—mainly members of the Puritan tradition—and were ejected from their positions in the Anglican Church. Many of them continued to preach, however, often under the protection of local gentry. These preachers became known as "Dissenters" or "Nonconformists."

In 1688, the passing of the Act of Toleration allowed freedom of worship for the Dissenters, who included Presbyterians and Quakers, meaning that (unlike Catholics at the time) they were no longer forced to worship in secret. But they did have to get licences for their places of worship, and they were banned from holding public office and—relevant to this story—from going to English universities. Nonconformist scholars and would-be ministers instead would go to Scottish universities, notably Edinburgh, or Dutch ones, in particular Leiden.

The Bayes family were Nonconformists. They were also wealthy—Richard Bayes, Thomas's great-grandfather, got rich in the Sheffield steel industry, making cutlery. Richard and his wife, Alice (née Chapman), had two sons. One, Samuel, went into the ministry, as many scions of rich families did, whether Nonconformist or Anglican. He was lucky enough to reach university age during the Commonwealth period, and was allowed to study at Trinity College, Cambridge, graduating in 1656. Samuel became a vicar in Northamptonshire, despite his Nonconformist beliefs, although he was among the two thousand clergy who refused to obey the Act of Uniformity in 1662 and was removed from his parish. The other son, Joshua, Thomas's grandfather, followed Richard into the family business.

The Bayeses appear to have committed quite seriously to the Non-conformist mission at this point. Joshua funded the building of a chapel in Sheffield, and his sons-in-law—he had four daughters and three sons, although two daughters and a son died in infancy—were the founder and minister of another one.

Joshua's second son, also Joshua, was born in 1671. He studied philosophy and divinity at a Dissenting academy in the north of England, which was forced to move repeatedly because of government harassment and persecution of Nonconformist academics. After that, he became a minister at various chapels in London, first in Southwark and then near Farringdon. According to Bellhouse, he was respected "both as a preacher and as a man of learning" by his flock.

He was also very much a classic Puritan family man, with a vast brood of children. He married his wife, Anne (née Carpenter), in October 1700, although the exact date is not known, likely because they were married in a Nonconformist chapel. Birth, death, and marriage registries were kept by the Church of England, while Nonconformist groups' records were often "kept secret, or not at all, for fear of religious discrimination."

For the same reason, the birth dates of Joshua and Anne's impressive tally of seven children are not known. All seven survived to adulthood, which was reasonably unusual at the time—about a third of English children born in the early eighteenth century died before the age of five.[4] We know that Thomas, the eldest, died in April 1761, aged fifty-nine, so was "with probability 0.8"[5] born in 1701 (alternatively, early 1702). His siblings were, in order of birth, Mary, John, Anne, Samuel, Rebecca, and Nathaniel; we know the years they died and how old they were when they did (John died youngest, age thirty-eight, in 1743, while Rebecca lived to eighty-two), but not their exact birth dates.

The family behaved as you'd expect a wealthy, educated family of the time to behave. One son, John, went to Lincoln's Inn and studied law, and was called to the bar in 1739. Samuel and Nathaniel went into trades, like their grandfather and great-grandfather—Samuel sold linen, while Nathaniel was a grocer. Mary never married; Anne and Rebecca married well-to-do men of their social station, a textile dealer and an attorney, respectively. And Thomas, of course, followed his father into Nonconformist ministry.

As a boy, Thomas was probably educated by a friend of the family, John Ward, later a professor of rhetoric at Gresham College, Cambridge, and a fellow of the Royal Society. Thomas's father helped pay for the printing of Ward's no doubt fascinating book *The Lives of the Professors of Gresham College*, and Ward's biographer says that he was "induced to undertake the education of a certain number of the children of his friends" and opened a school in Moorfields.[6] There is also a suggestion that he was educated as a boy by Abraham de Moivre, one of the great pioneers of probability theory, who had been forced to flee France for London and earn a living there as a tutor, although that appears to just be speculation.[7]

Thomas was a clever young man: a letter from Ward written in 1720, when Thomas would have been eighteen or nineteen, makes clear that Bayes could read Greek and Latin fluently—the letter is, after all, *in* Latin—although Ward had advice for how he could improve his Latin composition.

Despite his family's wealth and connections, and his own brains, as a Nonconformist Thomas was barred from the English universities. In 1719 he went up to Edinburgh, where he appears to have studied under Colin Drummond, professor of logic and metaphysics. The 1720 letter from Ward also tells us that Bayes studied mathematics, to Ward's satisfaction: "The order which you follow in the rest of your studies I cannot but highly approve of. In occupying yourself simultaneously with both mathematics and logic you will more clearly and easily notice what and how much each of these excellent instruments contributes to the directing of thought and sensation." [8]

But the main reason Bayes was in Edinburgh was to study divinity and prepare for his life as a minister. In 1720 he joined Divinity Hall, where records show that as part of his work, he submitted analyses of verses from the book of Matthew. The last is dated January 1722, so he must have stayed in Edinburgh at least until then.

The next thing we know about his life is that he turned up in London some time before 1728, when his name appears on a list of ministers submitted to a committee of Presbyterians, Independents, and Baptists—a committee of which Joshua, Thomas's father, was a frequent member and occasional chairman. Thomas at that point was an approved minister—he had the qualifications—but not yet in place at a chapel. By 1732, he had—according to that year's version of the list—joined his father at the chapel in Leather Lane, near Farringdon. By early 1734, he had moved to Tunbridge Wells in Kent, to take up a ministry of his own.

The nature of Bayes's belief is not exactly known. We know he was a Nonconformist, but that only narrows it down so far. But it does mean he probably had some very unusual, even flatly heretical, beliefs for his time.

He wasn't an Anglican. Nor was he a Catholic. The two doctrines are different, but not all that different—they differ on what seem to the outsider relatively small points. The Catholics believe salvation comes only through the Church, whereas the Anglicans believe that having faith in Jesus Christ and following his teachings get you to Heaven, even if you've never met a priest in your life. Catholics believe that the Eucharist wafer and wine literally become the body and blood of Christ in the Communion ceremony, whereas most Anglicans think it is merely imbued with his Spirit. They all, though, believe in the Holy Trinity—God the Father, God the Son, and God the Holy Spirit—and that God is both one substance and three persons.

Some of the Nonconformists had very different beliefs. In particular, Arians and Socinians denied the Trinity (and were viewed as heretics by mainstream Christians as a result). Arians believed that God the Father was the supreme God, and that Jesus, his son, was a lesser god who had always existed, even before he physically arrived on Earth. By contrast, Socinians agreed that Jesus was a lesser god, but believed he was brought into being only at the time of the Nativity. Later, Unitarianism grew out of those two heresies. It denied the Trinity too, but went further, saying that there is only one God, and that Jesus was not divine.

These beliefs grew fairly widespread among Presbyterian congregations in the eighteenth century. "The Presbyterians were really free thinkers," says Bellhouse, though not so free that these heretical beliefs didn't lead to tensions: in 1719, James Peirce and Joseph Hallett, two preachers, were expelled from Presbyterian churches in Exeter, having been accused of the Arian heresy.[9]

Bayes's first publication was a work of theology, *Divine benevolence: Or, an attempt to prove that the principal end of the divine providence and government is the happiness of his creatures: being an answer to a Pamphlet, entitled, Divine rectitude; or, An Inquiry concerning the Moral Perfections of the Deity. With a refutation of the notions therein advanced concerning beauty and order, the Reason of Punishment, and the Necessity of a State of Trial antecedent to perfect Happiness*, published in 1731.[10] His name was not on the author page (although, to be fair, there would hardly have been room), but it is widely accepted to be his work. His friend Richard Price refers to it in his own writings, and names Bayes as the author.

Divine Benevolence was a work of theodicy: an attempt to explain why God, if all-powerful and all-benevolent, allows evil in the world. As David Hume put it, apparently quoting Epicurus: "Is he willing to prevent evil, but not able? then is he impotent. Is he able, but not willing? then is he malevolent. Is he both able and willing? whence then is evil?"[11]

Bayes was responding to a tract by John Balguy, an Anglican theologian, who argued that the suffering in the world was caused because God's goodness was about doing what is "right and fit," which is not necessarily what we humans enjoy.[12] Bayes, by contrast, believed that God is indeed *benevolent*, and wants us to be happy. Since a lot of us *aren't* happy, much of Bayes's argument was spent explaining why God might not try to make us happy, even though he can and wants to. It was, apparently, quite controversial and widely read.

But *Divine Benevolence* doesn't go into Bayes's own faith. Bayes's father, Joshua, was a "moderate Calvinist who was tolerant of a variety of views,"[13] but Bellhouse argues that Thomas was probably an Arian or a Socinian, "halfway to being a Unitarian." "He was not your run-of-the-mill orthodox Christian," says Bellhouse. "He trained as a Presbyterian minister, but he was probably a Socinian."

The clue is the company he kept. He was friends with one James Foster, another Dissenting minister, who was himself friends with the two Exeter ministers who'd been expelled for Arianism. Foster had also written a pamphlet, *An Essay on Fundamentals in Religion*,[14] arguing that the Trinity was not essential to Christianity, which sounds dangerously heretical to me.

William Whiston, Isaac Newton's successor as the Lucasian professor of mathematics at Cambridge, was another associate of Bayes, and at one breakfast the two men had together he asked Bayes whether the sermon at the local Anglican church that weekend would include the Creed of Athanasius, which lays out the doctrine of the Trinity. Whiston said he would leave the service if so, and Bayes reassured him it would likely not.

Bayes would also, upon his death, leave £200 ($242) to John Hoyle and Richard Price, two Nonconformist ministers in London, both of whom were Arian in their faith and both of whose churches later became Unitarian. Price in particular was a close friend—when Bayes died, it was Price who reworked and published the famous essay that contained Bayes' theorem.

Thomas Bayes lived in a high-society world. His peers tended to be university-educated, often with doctorates of divinity, and many of them were members of the nobility.[15] You can see this from his associations with well-respected figures like Ward and Whiston. At Tunbridge Wells, Bayes continued to mingle with well-known or well-connected people. The most important appears to have been Philip Stanhope, the 2nd Earl Stanhope.

Tunbridge Wells in those days was "chiefly a tourist town."[16] It was reachable within a day by carriage from London, and its most notable feature was a large and much-admired spa, fed by a local spring. Stanhope, who became the earl at the age of seven after the death of his father, and whose family home of Chevening was just a few miles from Tunbridge, was a regular attendee there from his early twenties. He was younger than Bayes, born in 1713.

The young Earl was an enthusiastic amateur mathematician. As a child, his uncle and guardian had attempted to push him away from math and toward the literary arts, but once he reached the age of majority he took it up with a will. "He has read a good deal of Divinity, Metaphysicks, and Mathematicks," wrote a contemporary.[17] "He is always making mathematical scratches in his pocket-book, so that one half the people took him for a conjuror, and the other half for a fool," wrote another.[18]

Stanhope appears to have built a network of fellow scientists and mathematicians. As well as Bayes, this included Robert Smith, a University of Glasgow mathematician, whose works Stanhope had published posthumously; Joseph Priestley, the chemist and discoverer of oxygen;

and John Eames, a theologian-scientist and friend of Isaac Newton. All of them, and many others in Stanhope's network, were Nonconformists of one kind or another, and most of them were amateurs—gentleman scientists, hobbyists.

"He didn't seem like a modern academic," Bellhouse says of Bayes. "He was more of an amateur, a virtuoso. He did it for his own pleasure rather than having a research agenda."

So Stanhope, and Bayes, clever men of considerable leisure and undemanding jobs, made hobbies of mathematics. "What the rich did in the eighteenth century was to get involved in science," Bellhouse says. "It's similar to rich people nowadays getting involved in sports teams."

The two men wrote to each other regularly; the correspondence was found relatively recently among Stanhope's effects. It appears that Stanhope met Bayes in the 1730s, having either recently obtained a copy of Bayes's paper *An Introduction to the Doctrine of Fluxions*,[19] or being given it shortly after.

Fluxions was a defense of Newton's calculus against an attack by the philosopher George Berkeley. Bayes was a committed supporter of Newton. "Some [Nonconformists] were hesitant to teach mathematics," says Bellhouse, "in case it led to Newtonian science, and from there to atheism. But a much larger group among the Nonconformists said that it's important to study mathematics—you need to understand God's universe."[20]

Berkeley argued that Newton had made, in essence, a divide-by-zero error: that one of his terms in a key equation was simultaneously zero and non-zero, and that his "doctrine of fluxions" was therefore inherently contradictory. Bayes, in his response, tried to firm up Newton's definitions more rigorously, establishing exactly what various terms meant.

After that, Bayes did some work on infinite series and their relationship to derivatives. A *derivative* is the rate of change of a slope on a graph. If you have a graph of time (seconds) and distance (meters), the shape of the line tells you something about the speed (meters per second). If the line is straight, your speed is constant. If the line is curved, your speed is changing. A derivative measures the slope of the curve at an exact point, so you are able to work out the speed for any given distance or time. And you can go up a layer: divide your speed by your time and find your acceleration, which is the *second derivative* of distance and time.

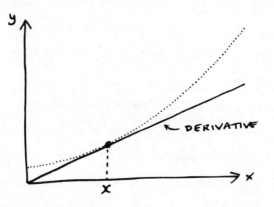

An *infinite series*, meanwhile, is just a mathematical series that goes on forever. If I say "x equals one plus two plus three plus four and so on,"[*] then that's an infinite series, and x is equal to infinity. But some infinite series do not equal infinity. For instance, if I say "x equals a half

---

[*] In notation: $x = 1 + 2 + 3 + 4 + 5 \ldots$

plus a quarter plus an eighth plus a sixteenth plus a thirty-second and so on,"* then that's an infinite series as well, and x is equal to one.

Bayes showed that the derivative of a number y is equal to an infinite series of y at time T minus half of y at time T + 1 plus a third of y at time T + 2 and so on. It's a neat little theorem, found in Stanhope's papers long after both men were dead ("Theorem mentioned to me at Tunbridge Wells by Mr Bayes Aug. 12. 1747,"[21] says a laconic note on a scrap of paper) and which, Bellhouse believes, was not independently discovered until a quarter of a century later by the French mathematician Joseph-Louis Lagrange.[22]

It was around this time that Bayes grew interested in probability theory. But before we get into that, we need to turn to the history of the mathematics of chance, and what people were working on at that point.

PASCAL AND FERMAT

Traditionally, the story of the study of probability begins in French gambling houses in the mid-seventeenth century. But we can start it earlier than that.

The Italian polymath Gerolamo Cardano had attempted to quantify the maths of dice gambling in the sixteenth century. What, for instance, would the odds be of rolling a six on four rolls of a die, or a double six on twenty-four rolls of a pair of dice?

---

*In notation: $x = (\frac{1}{2}) + (\frac{1}{4}) + (\frac{1}{8}) + (\frac{1}{16}) + (\frac{1}{32}) \ldots$

His working went like this. The probability of rolling a six is one in six, or ⅙, or about 17 percent. Normally, in probability, we don't give a figure as a percentage, but as a number between zero and one, which we call p. So the probability of rolling a six is p = 0.17. (Actually, 0.1666666 . . . but I'm rounding it off.)

Cardano, reasonably enough, assumed that if you roll the die four times, your probability is four times as high: ⁴⁄₆, or about 0.67. But if you stop and think about it for a moment, that can't be right, because it would imply that if you rolled the die six times, your chance of getting a six would be one-sixth times six, or one: that is, certainty. But obviously it's possible to roll six times and have none of the dice come up six.

What threw Cardano is that *the average number of sixes* you'll see on four dice is 0.67. But sometimes you'll see three, sometimes you'll see none. The odds of seeing a six (or, separately, at least one six) are different.

In the case of the one die rolled four times, you'd get it badly wrong—the real answer is about 0.52, not 0.67—but you'd still be right to bet, at even odds, on a six coming up. If you used Cardano's reasoning for the second question, though, about how often you'd see a double six on twenty-four rolls, it would lead you seriously astray in a gambling house. His math would suggest that, since a double six comes up one time in thirty-six (p ≈ 0.03), then rolling the dice twenty-four times would give you twenty-four times that probability, twenty-four in thirty-six or two-thirds (p ≈ 0.67, again).

This time, though, his reasonable but misguided thinking would put you on the wrong side of the bet. The probability of seeing a double six in twenty-four rolls is 0.49, slightly less than half. You'd lose money betting on it. What's gone wrong?

A century or so later, in 1654, Antoine Gombaud, a gambler and amateur philosopher who called himself the Chevalier de Méré, was interested in the same questions, for obvious professional reasons. He had noticed exactly what we've just said: that betting that you'll see at least one six in four rolls of a die will make you money, whereas betting that you'll see at least one double six in twenty-four rolls of two dice will not.

Gombaud, through simple empirical observation, had got to a much more realistic position than Cardano. But he was confused. Why were the two outcomes different? After all, six is to four as thirty-six is to twenty-four. He recruited a friend, the mathematician Pierre de Carcavi, but together they were unable to work it out. So they asked a mutual friend, the great mathematician Blaise Pascal.[23]

The solution to this problem isn't actually that complicated. Cardano had got it exactly backward: the idea is not to look at the chances that something *would* happen by the number of goes you take, but to look at the chances it *wouldn't* happen.

In the case of the four rolls of a single die, your chance of *not* seeing a six on any one throw is ⅚, or $p \approx 0.83$. If you roll it again, your chance of not seeing a six on either throw is 0.83 times 0.83, or just shy of 0.7. Each time you roll the die, you *reduce* the chance of *not* seeing a six by 17 percent.

If you roll the die four times, your chance of *not* seeing a six is 0.83 × 0.83 × 0.83 × 0.83 ≈ 0.48. (To save time, we can say "0.83 to the power 4," or "0.83 ^ 4.") So your chance of *seeing* a six is 1 minus 0.48, or 0.52, or 52 percent. If you bet at even odds one hundred times, you'd expect to win fifty-two times, and you'd be in profit.

But look what happens when we do it with the two dice, looking for a double six. Your chance of seeing a double six on one roll of two dice is $\frac{1}{36}$, or $p \approx 0.03$, as we said earlier. So your chance of *not* seeing a double six is $\frac{35}{36}$ or about 0.97.

If you roll your dice twenty-four times, your chance of *not* seeing a double six is 0.97 multiplied by itself twenty-four times (0.97 ^ 24). If you do that sum, you end up with 0.51. So the chance of *seeing* a double six is 0.49. If you bet at even odds, you'd expect to see it forty-nine times in a hundred, and you'd lose money.

(We should take a moment, here, to recognize the absolutely heroic amount of gambling that Gombaud must have been doing in order to be able to tell that his 52 percent bet was coming off, but his 49 per-cent bet wasn't. Apparently, he had deduced, correctly, that you need twenty-five rolls of the dice, not twenty-four, for it to be a good bet. Gombaud was a man who enjoyed his dice-rolling.)

This led Gombaud to raise another question with Pascal. Imagine two people are playing a game of chance—cards or dice. Their game is interrupted halfway through, with one player in the lead. What's the fairest way to divide the pot? It seems wrong to simply split it down the middle, since one person is winning; but it's also unfair to give it all to the player in the lead, since they haven't actually won yet.

Pascal found this fascinating, and exchanged a series of letters[24] dis-cussing the problem with his contemporary, Pierre de Fermat, of Last Theorem fame.

Again, this problem goes back a few centuries. The Italian monk Pacioli had a go at solving something like it in 1494, in his work *Summa de arithmetica, geometrica, proportioni et proportionalità.*[25]

He imagines that two players are playing a ball game in which you win ten points for each goal, and the winner is the first person to get to sixty points.* One of the players has reached fifty points, and the other has reached twenty, before the game is interrupted. How should the winnings be split?

Pacioli reasons that, since one player has scored five-sevenths of all the points so far scored, that player should win five-sevenths of the pot.

Forty-five years later, the aforementioned Cardano—he who'd got the math backward on the dice problem, so could perhaps have shown a little more humility—scoffed that Pacioli's solution was "absurd." He imagined a slightly different scenario, where two players play a game of first to ten. One has seven points, and one has nine. In that situation, by Pacioli's system, the first player should get nearly half the pot—seven-sixteenths—and the second player only slightly more, nine-sixteenths. But that seems obviously unfair, since one player only needs one point to win, while the other needs three.

Cardano suggested a better route. "His major insight," writes Prakash Gorroochurn, "was that the division of stakes should depend

---

*As an aside, this drives me mad. Why sixty and ten? Why not six and one? The Quidditch scoring system in Harry Potter is equally stupid—the two means of scoring get you 10 and 150 points, respectively. Why not 1 and 15? Why the extra zero? While I'm ranting about Quidditch, it's also completely insane that grabbing the snitch is worth fifteen times scoring a goal and essentially renders the entire effort of the team to score goals pointless.

on how many rounds each player had yet to win, not how many rounds they had already won."[26]

But Cardano didn't get all the way there. He suggests using the ratio of the "progressions" of the two players' still-required scores. The progression of a number, in his jargon, is that number, plus that number minus one, plus that number minus two, and so on down to one. So the progression of five would be $5 + 4 + 3 + 2 + 1 = 15$.

In the example Cardano gave, the first player has three points still to win. The progression of three is six ($3 + 2 + 1 = 6$). The second player has one point still to win, and the progression of one is one ($1 = 1$). So, for Cardano, the pot should be divided six parts to one in favor of the second player.

This is better than Pacioli's system, or at least gets you closer to the true answer. But it's still wrong.

This is where Pascal and Fermat come into the picture. They realized the key point: It's not how close to the finish you are, or how far from the start you've come, that matters. It's *the number of possible outcomes that remain*, and how many of those outcomes favor one player over the other.

Pascal, in a letter to Fermat, imagined a simple situation: two gamblers are playing a game of first to three points. They have each bet thirty-two pistoles (a gold coin used in currency at the time), so the total pot is sixty-four pistoles.

Let's say it's all square at two points each, and they suddenly have to end the game. In that case, reasons Pascal, it's easy enough to divide. You just split it in half, thirty-two each.

But what if they'd had to end it one turn before, when one player had two points and the other player had one? Pascal extends the reasoning. They would have split it evenly had it gone to two rolls each, so the first player is sure of at least half the pot—even if that player were to lose the next throw, they would still have that. The other half is still a going concern. "Perhaps I will have them and perhaps you will have them," Pascal imagines the first player saying. "The risk is equal. Therefore let us divide the thirty-two pistoles in half, and give the thirty-two of which I am certain besides." So the first player will take 32 + 16 = 48, or three-quarters of the pot.

Another way to look at it is to say that there are four possible ways the game could have gone, had it continued. Player One could have won the first throw and the second; they could have won the first throw but lost the second; they could have lost the first throw but won the second; and they could have lost the first throw and lost the second.

Only in the fourth scenario does Player Two win the pot. If Player One wins the first throw, the second throw is irrelevant: Player One has made it to three points. So half the outcomes are wins for Player One without even going to the last throw. And even if they lose that first throw, they're still in with a fifty-fifty chance of winning.

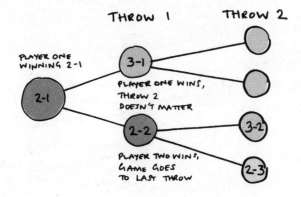

So the fair distribution of the pot, if the two players have to stop playing with one player up two to one, is three to one, just as Pascal said.

You can expand this, and Pascal does. Imagine that Player One was winning two–nil, not 2–1. If they win the next throw, they win. But if they lose, the other player is back to 2–1. And we've just seen that, from that point, their chance of winning the pot is 75 percent. In Pascal's example, Player One would say: "If I win, I shall gain all; that is sixty-four. If I lose, forty-eight will legitimately belong to me. Therefore give me the forty-eight that are certain to be mine, even if I lose, and let us divide the other sixteen in half, because there is as much chance that you will gain them as that I will."

So now Player One has a seven-eighths, or 87.5 percent, chance of winning, so the fair division is that Player One takes fifty-six pistoles out of sixty-four. Again, you can draw this as a diagram:

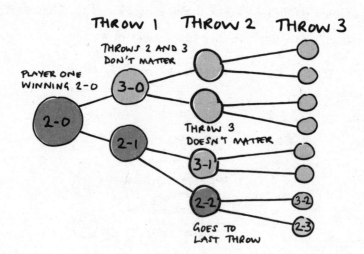

But how about if Player One only has *one* point, and Player Two zero? Then you extend it one further back, said Pascal. If Player Two wins the first throw, then it's one–all, and an equal chance of winning. But if Player One wins the first throw, then it's two–nil, and we know the situation: she has seven-eighths chance. Out of a possible sixteen outcomes, Player One wins in eleven, so she should win eleven-sixteenths of sixty-four pistoles, or forty-four.

This is the great insight of probability theory: that we should look at the possible outcomes from a given situation, not what has gone before. But laboriously counting out the number of possible outcomes as we have above takes quite a long time, so Pascal and Fermat worked on ways of making it quicker.

You can work it out as a sum, but it's complicated if you have large numbers of rounds left to play. You need to work out the maximum possible number of remaining throws—that is, the number Player One needs to win, plus the number Player Two needs to win, minus one. If someone's one–nil up in a first-to-three game, that's four. (The highest score the game could reach is 3–2, five points in total.) Four remaining rounds means sixteen remaining possible outcomes—that is, two times itself four times. And *then* you need to work out which of the outcomes correlate to a win for Player One, which involves a lot of superscript and Greek letters and would just tire us all out.

Luckily, Pascal came up with a cheat. He wasn't the first to use what we now call Pascal's triangle—it was known in ancient China, where it is named after the mathematician Yang Hui, and in second-century India. But Pascal was the first to use it in problems of probability. It looks like this:

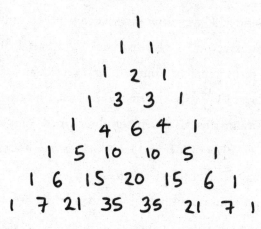

It starts with 1 at the top, and fills out each layer below with a simple rule: on every row, add the number above and to the left to the number above and to the right. If there is no number in one of those places, treat it as zero.

Pascal realized that he could use the triangle to solve the problem of points. Take our example. There are a maximum of four rounds left to play, so you count down four rows from the top (counting the very top row, the solitary 1, as row zero). Player One needs two more points to win, so take off the first two numbers from the left. Add the remaining numbers together, divide them by the total value of that row, and you get your chance of winning.

In this case, count down four rows from the 1, and you find you're on a row that goes: 1 4 6 4 1. Take the first two numbers away, and you're left with 6 4 1, which add up to 11. The whole row adds up to 16. That is, a 11⁄16, or a 68.75 percent chance: $p = 0.6875$.

Try it for the other examples we've looked at. If Player One has 2 points and Player Two has 1, then there are a maximum of two possible goes left, and Player One only needs to win one of them. So you count down two rows to the 1 2 1 row, you remove the 1, and you're left with ¾, or p = 0.75. It's astonishingly neat, and saves you lots of time.

It works for any event that has two equally likely outcomes, like coin-flipping or games between equally matched opponents. For a given number of goes, X, you look at row X (again, with the very top line being row zero). That gives you the total number of possible outcomes. So if you flipped a coin seven times, you'd count down to row seven, the one starting 1 7 21, add those outcomes up, and you find that it equals 128. So there are 128 possible outcomes.

Now, if you want to know what the possibility is of seeing *exactly* Y outcomes, say heads, on those seven flips:

It's possible that you'll see no heads at all. But it requires every single coin coming up tails. Of all the possible combinations of heads and tails that could come up, only one—tails on every single coin—gives you seven heads and zero tails.

There are seven combinations that give you one head and six tails. Of the seven coins, one needs to come up heads, but it doesn't matter which one. There are twenty-one ways of getting two heads. (I won't enumerate them all here; I'm afraid you're going to have to trust me, or check.) And thirty-five of getting three.

You see the pattern? 1 7 21 35—it's row seven of the triangle.

So if you want to know the chance of getting exactly Y heads on X flips, you count down X rows from row zero and look at the number that's Y from the left (again, counting the 1 at the left as 0). Then you divide that second number by the first. Say you want to know the odds of getting exactly five heads, you look at row seven—that's the 1 7 21 35 35 21 7 1—and starting from zero, you count five along. That's the second twenty-one. So $2\frac{1}{128} \approx 0.164$, or about a one-in-six chance.

To find the chance of getting *at least* five heads, you just add the number of possible ways of getting six heads or seven heads to the ways of getting five heads: 21 + 7 + 1 = 29. Then you divide it by 128 as we did before. That's what Pascal was doing to work out the fairest way to split the pot.

Pascal's triangle is only one way of working out the probability of seeing some number of outcomes, although it's a very neat way. In situations where there are two possible outcomes, like flipping a coin, it's called a "binomial distribution."

But the point is that when you're trying to work out how likely something is, what we need to talk about is the number of outcomes—the number of outcomes that result in whatever it is you're talking about, and the total number of possible outcomes. This was, I think it's fair to say, the first real formalization of the idea of "probability."

THE LAW OF LARGE NUMBERS

Pascal and Fermat's letters were the beginning of the modern idea of *probability theory*, although in its early days it was known as *the doctrine of chances*. You can think of the underlying idea as being that the probability of some event is the number of ways that event can occur, divided by the total number of things that could occur.

Jacob Bernoulli, a Swiss mathematician, introduced the next stage.

To use one of the examples we just looked at, if you do flip seven coins 128 times, it's very unlikely that you'll see zero heads once, one head seven times, two heads twenty-one times, and so on.

But if you flipped the coins 128 *million* times, you'd very probably see zero heads something like a million times, one head 7 million times, two heads 21 million times, etc. Or, in a more basic demonstration, if you flipped a fair coin twice, you might well not see one head and one tail—in fact, 50 percent of the time, you wouldn't; you'd see two heads or two tails. But if you flipped it a million times, you'd probably see somewhere around half of them come up heads and half come up tails.

Bernoulli's claim was that the more you flipped the coin, the closer, on average, to the "true" probability your results would be.

You might reasonably say that this is pretty obvious. And also, so what? You know you'll see roughly half of the coin flips come up heads. You don't need to flip the coin a million times to prove it.

But so far, we've only been looking at the probability of known events in games of chance—dice-rolling, coin-flipping. Things where we know (in theory, at least) the probability of the basic components of the game in advance. It's axiomatic that a coin flip is a fifty-fifty shot, that when we roll the dice, we'll see a one once in every six throws.

Sometimes, though, we might wonder: Is the coin fair? Are the dice loaded? *When can we tell?* Or maybe we aren't playing dice at all; maybe we're trying to work out how the world actually is, how often things happen. We have to go outside games, where everything is established in the rules, and go out into the real world of messy chance and uncertainty.

Bernoulli lived in Switzerland in the seventeenth century, one of a family of genius mathematicians. (It is important not to get confused between "Bernoulli's theorem," which we are about to discuss, named for Jacob, and "Bernoulli's principle," something completely other, named for his nephew Daniel. There were also three Johanns, two Nicolauses, and a second Jacob among the Bernoullis of the sixteenth and seventeenth centuries whose names are blue on Wikipedia.)

What he was interested in was not just in fact games of chance. He was also interested in balls in urns.

Imagine the following situation.[27] You are presented with a large urn. Inside the urn are a number of black and white balls. But you don't know the ratio of the two colors. You draw some balls out, and you get some black ones and some white ones. Say you draw five balls, and you get three black balls and two white balls. Using that result, what can you say about the contents of the urn as a whole?

Now we're no longer talking about working out the probability of seeing some result, given certain facts about the world. We're talking about the exact opposite: What is the probability of the world being a certain way, given the results that we're seeing? The two ideas are *sampling probabilities*—what can we predict about a sample of something, given what we know about the whole?—and *inferential probabilities*—what can we know about the whole, given a sample we've taken?

I'm just going to stop here for a moment and really drill this home. This distinction is *crucial*. It doesn't sound like much, maybe, but it's the whole game. This is what modern statisticians—modern scientists—do all day. They don't sit around working out the probabilities of drawing a straight flush in Texas Hold'em. That's a straightforward thing to work out, if you know how many cards there are in the deck; any high school math student could do it. They don't worry about how likely it is you'll roll five or more sixes on twenty dice. You can do that with Pascal's triangle in a few seconds. What they do is try to establish what the data tells us about a hypothesis. If I give five hundred people a COVID vaccine, and five hundred people a placebo, and then ten people get COVID in the placebo group and only one in the vaccine group, what does that tell us? How sure can we be that the vaccine works?

This is what Bernoulli was trying to do. And his solution was brilliant and insightful, and—at least according to Aubrey Clayton, the author of *Bernoulli's Fallacy: Statistical Illogic and the Crisis of Modern Science*—wrong. For Clayton, and the school of statistical thinking he represents, Bernoulli also, sadly and inadvertently, set statistical thinking down the wrong path for the next three centuries. Whether Clayton is right or not is the subject of more than a century of bitter academic dispute, which we will discuss elsewhere in this book. But first, let's look at what Bernoulli did, and why.

Bernoulli wanted to know how confident we could be in the contents of an urn after we had drawn some number of balls. Say you've got an urn with balls in it. Each time you draw a ball, you put it back and shake the urn up again.[28] (That's important, so that the chance of drawing a black or a white ball remains the same throughout.) The balls are well mixed and equal in size and weight; you've got no way of telling whether a ball is white or black before you pick it, and there's no reason for black or white balls to tend to be higher or lower in the urn. If you draw X balls out of the urn, and Y of them are white, what can you say about the ratio of white balls to black in the urn?

The larger your sample, the more likely it is to be close to the true ratio. If the real ratio in the urn is that three out of every five balls are white, and you draw five balls, it's not that likely that you'll see exactly three whites and two blacks. But if you draw fifty balls, you might not see exactly thirty and twenty, but it's much more likely that you'll see something close to it. Bernoulli himself recognized that it was something that "even the stupidest man knows by some instinct of nature per se and by no previous instruction."[29] (And, indeed, in 1951 it was shown that even quite young children grasp this concept intuitively.)[30]

But Bernoulli wanted to take it further than that. His insight was that there are three components to this question: how big a sample you take, how close to the true answer you need to be, and how confident in your answer you need to be. He realized that you can never be *truly* certain of the actual ratio. What you can have instead, he said, is "moral certainty"—that is, a given degree of confidence in a given spread of results.

So you might want your sample to give you a result that is 99 percent likely to be within 1 percent of the true value. Or you might want it to be 70 percent likely to be within 10 percent of the true value. Bernoulli proved that, for either of those, or any other combination, there is a number of balls that you can draw out of the urn that will give you that level of confidence. He also showed that there is no point at which either you reach certainty or that increasing your sample ceases to give you greater confidence.

To express it as a mathematical theorem (these are not Bernoulli's own words, but a modern rephrasing): "[We] can always specify the number of observations *n* such that, for any probability we wish, the absolute difference between the sample proportion m/n (where m is the number of positive cases) and the true proportion p is less than or equal to some number $\in$ of our choosing."[31]

Note that there are three moving parts and adjusting any one of them means changing at least one of the others. So if you've drawn a sample that's large enough for you to be 90 percent sure that it's within 10 percent of the true answer, but you want to be 99 percent sure, you either have to adjust your spread—make it wider than 10 percent—or you have to take a larger sample. (As Aubrey Clayton points out, it's like the project-management mantra "Fast, good, or cheap. Pick two." In this case, it's "Precise estimates, high certainty, or small samples. Pick two.")[32]

Having proved that this was the case, Bernoulli wanted to put numbers on it. Exactly what level of confidence—what degree of "moral certainty," in his words—could you achieve with a given sample size? And he managed it: if the true number of balls in the urn was 3,000 white and 2,000 black, Bernoulli showed that with a sample size of 25,500, you would get a result that lies within 2 percent of the answer 999 times out of a thousand.

(Which is an inconveniently large sample size for someone working in early modern Europe without the benefit of a computer or ready access to psych undergraduates willing to take part in social science research for beer money. It was, as Stephen Stigler points out in his *The History of Statistics*, larger than the contemporary population of Basel, Switzerland, where Bernoulli lived, and "more than astronomical: for all practical purposes it was infinite." The abrupt ending of *Ars Conjec-*

tandi after that line, says Stigler, suggests that "Bernoulli literally quit when he saw the number 25,500, mustering strength only to add one further sentence.")[33]

With more modern methods it would be possible to achieve Bernoulli's desired degree of certainty with a smaller sample size, but nonetheless, by today's standards, he was being extremely demanding of himself. We'll talk more about p-values and confidence intervals a bit later, but Bernoulli's preferred level of moral certainty—landing within a given distance of the target 999 times in every 1,000—is equivalent to a false positive rate of 0.001. Most social science, at least, asks for a false positive rate of 0.05—fifty times less stringent—although in some other sciences, notably physics, a higher standard is used.

What Bernoulli recognized, though, was that this wasn't only relevant to parlor games and gambling houses. With Bernoulli, we deal with probabilities all the time—he gives examples of trying to establish who committed a murder, or whether a document is fraudulent. This was all part of Bernoulli's wider aim: to create a philosophically robust way of using empirical evidence. For two thousand years, philosophers had argued whether the true route to understanding was with one's reason or with the senses. Plato argued that there was a true underlying reality to the universe—what he called Forms—but that our senses were untrustworthy and could never give us certain knowledge.[34] Plato said, therefore, that the route to understanding was by reasoning, not experiment.

Bernoulli was a physicist and experimentalist. He accepted that we could never know anything with absolute certainty. But, he said, that doesn't have to mean that all things are equally likely. If we roll a die one hundred times and it comes up six every time, we can't say with absolute certainty that it's loaded. But we can say it's very likely that it is. In a foreshadowing of things we will discuss later—the idea of probability, and specifically Bayes' theorem, as an extension of formal logic—Bernoulli thought that we could talk about certainty as a number: 1 for complete certainty, 0 for complete impossibility.[35] And that meant that you could have degrees of certainty, and improve that certainty by experiment.

The trouble, for Clayton at least, was that Bernoulli was still talking about *sampling probabilities*, not *inferential probabilities*. Or rather—he didn't draw a distinction between the two. Bernoulli had successfully shown that the ratio of white balls to black balls in the sample will probably be close to the true ratio in the urn as a whole (exactly how close probably depending on the sample size). What he *assumed* was that this therefore meant that it was equally likely that the true ratio of white balls to black balls in the urn would be close to the ratio in the sample. But in that he was wrong; the probabilities can be very different indeed. It was not until the Reverend Thomas Bayes that it was understood how Bernoulli was wrong.

DE MOIVRE ON THE NORMAL DISTRIBUTION

Abraham de Moivre was a French Protestant who fled persecution by the Catholic authorities in his hometown of Vitry after being imprisoned for two years.[36] In 1688, at the age of twenty-one, he arrived in London, where he read Newton and became a tutor, learning mathematics as he did so. He took Bernoulli's ideas one stage further.

Think back to Fermat and Pascal working out how you ought to divide a pot if a game was ended early. They worked it out by looking at how likely each player was to win, given where the game stood when they ended it. And that came down to how many of the remaining outcomes led to a win for Player A as opposed to Player B.

What they were discussing was what's now known as a *binomial distribution*. If you flip a coin, it can come up either heads or tails. If you flip a coin twice, it will come up either heads then heads, heads then tails, tails then heads, or tails then tails. So there's only one way of getting two heads or two tails, but two ways of getting one heads and one tails. You can write the probability out:

NUMBER OF HEADS	PROBABILITY
0	1/4
1	1/2
2	1/4

Or you can draw it on a graph:

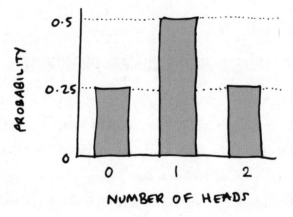

That's the distribution of outcomes for two coin flips. (Or any other event with two equally likely outcomes.) For four coins:

NUMBER OF HEADS	PROBABILITY
0	1/16
1	4/16 (1/4)
2	6/16 (3/8)
3	4/16 (1/4)
4	1/16

(You might notice the Pascal's triangle numbers in there again!) And as a graph:

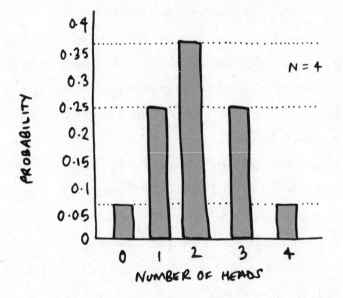

You can calculate the probability of getting any given number of heads (x) from a given number of coin flips (N) with an equation that I won't write here (use an online calculator, there are loads), but that starts by finding the factorials of the number of throws, the number of heads you want, and the number of throws *minus* the number of heads.

If you've not heard the term, a factorial of a number equals that number multiplied by that number minus one, multiplied by that number minus two, and so on, down to one. So the factorial of five is 5 × 4 × 3 × 2 × 1 = 120. Doing that for large numbers is, in technical terms, a gigantic pain in the arse. (They get bigger extremely quickly. The factorial of six is 720. The factorial of ten is 3,628,800.)

And you're often not just interested in the chance of getting *exactly* x heads. If we imagine a gambling situation again, and someone says, "I'll bet you £50 to £10 that you won't get sixty or more heads out of a hundred coin flips," is that a good bet? Well, using the binomial distribution calculation, you'd have to work out the factorials of sixty, forty, and one hundred, then plug that into the equation. Then you'd have to do it again, except with sixty-one and thirty-nine. And again with sixty-two and thirty-eight. And so on. For ages. Bernoulli actually *did* this stuff, which may be why his book took twenty years to write and he never actually finished it.

Of course, once someone's worked out the factorial of, say, 253—which is 507 digits long and ends with 62, apparently—they can write it down, and everyone else can use it without having to work it out again. But even taking that into account, it's a long and boring process.

What de Moivre noticed was the shape of the curve.[37] Look at the two graphs above: they both have a bulge in the middle and flattened edges. But in the N = 4 graph it's smoother and more noticeable.

If you do a larger number of coin flips, the curve becomes clearer still. For N = 12:

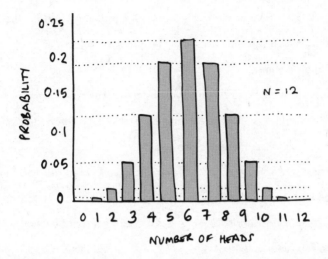

What de Moivre thought was that, instead of slogging through the equations to work out the odds of seeing sixty heads on a hundred coins, you could work out the mathematical expression for the curve, and then use the shape of the curve to get the probability of any outcome. The curve is what we now call the "normal distribution" or the "bell curve" (although statisticians I've spoken to dislike that latter term, because it doesn't actually look much like a bell).

IN CASE YOU NEED A REFRESHER
ON WHAT STANDARD DEVIATIONS ARE

We would now talk about what de Moivre was describing in terms of "means" and "standard deviations," a term that was not coined for another century and a half. I imagine most readers will know what means are (yer bog-standard average), but "standard deviations" is one of those terms that get thrown around a lot by proper mathematicians as though we'll all automatically know what it means, and I suspect most of us don't. The standard deviation is a measure of how spread out your data is around the mean.

Imagine you have three kids and you want to know how tall they are, on average. You measure them all, add it up, divide by three, and you get 160 cm. That's your mean, your average.

OK, fine, except there are lots of ways that you could get 160 cm. It could be that every kid is exactly 160 cm tall. Or that one is 157 cm, one is 160 cm, and one is 163 cm. Or it could be that two are little 130-cm-tall six-year-olds and one is 2.2 meters tall and plays college basketball. Or an infinite number of other combinations.

The key difference is how much they vary from the mean: the variance. Once you have the variance, you can easily work out the standard deviation by finding the square root of that variance.

You work out the variance by taking each child's height and subtracting your mean (in this case, 160 cm) from it. Then you square that number—that is, multiply it by itself. (If you don't do that, some of the numbers will be negative.) Then you take the average of those numbers.

Let's look at the example where one child is 157 cm, one is 160 cm, and one is 163 cm. Subtracting your mean from each of those gets you—3 cm, 0 cm, and 3 cm. Square them and you get 9 cm, 0 cm, and 9 cm. The average is 6 cm,* so the variance is 6. The square root of 6 is roughly 2.4, so that's your standard deviation.

In the case of the basketball player and the eight-year-olds, sub-

---

*9 + 9 + 0 = 18, 18/3 = 6.

tracting the mean from their heights gives you −30 cm, −30 cm, and 60 cm. The squares of those are 900, 900, and 3,600; the average of those is 1,800, so that's your variance. The square root of 1,800 is 42.4.

Once you've got your standard deviation, you can start talking about how far each value is from the mean in terms of standard deviation (which is usually written as SD, or with the Greek letter sigma, σ).

Take the eight-year-olds-and-the-basketball-player example. The SD is 42.4. The two eight-year-olds are 30 cm from the mean, so they're 30/42.4 = 0.7SD below the mean. The basketball player is 60 cm from the mean, so he's 1.4SD above the mean.

What's interesting here is that with normally distributed data and a sufficiently large sample, you can reliably predict what percentage of results will fall within a given distance of the mean, as measured in SD. In general, 68 percent of all values will be within 1SD of the mean—so if you're 1SD above the mean in height, then you're taller than about 84 percent of the population. Meanwhile, 95 percent of the population will be within 2SD, and 99.7 percent will fall within 3.

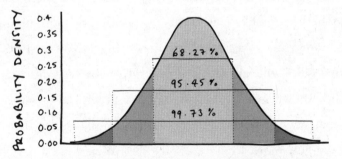

68% OF THE DATA IS WITHIN ONE STANDARD DEVIATION
95% OF THE DATA IS WITHIN TWO STANDARD DEVIATIONS
99.7% OF THE DATA IS WITHIN THREE STANDARD DEVIATIONS

De Moivre was able to show that, by working out the shape of the curve of the normal distribution (although it wasn't at that time *called* the normal distribution), you could get a quick approximation of the chance of seeing any given result. His methods said that 68.2688 percent of results fall within 1SD of the mean; the true answer is 68.2689 percent. For 2SD, his answer was 95.428 percent chance, compared to a real answer of 95.45 percent. For 3SD, it's 99.874 percent versus 99.73 percent.[38] (They also weren't called standard deviations, but he did use the idea and recognize that it was the scale by which deviations from the mean should be judged.)

So if you want to know how likely you are to see some result a given distance from the mean, what you need to do is work out the standard deviation of your data, and plug it into de Moivre's curve-working-out equation. You won't have to spend several days working out the factorial of 3,600.

What de Moivre also realized, which was an extension of what Bernoulli had been thinking about, was that the accuracy of your data—the size of your standard deviation—depends on how many samples you take. Bernoulli had spent twenty years tediously working out the required sample size for a single example—being 999 out of 1,000 sure that your results will fall within 2 percent of the true outcome. De Moivre had shown how to do it for any result, with impressive if not perfect accuracy—and where Bernoulli had simply shown that larger samples *did* give more accurate results, de Moivre took it a step further and quantified it. He showed that the accuracy of your estimate grew in proportion to the square root of the sample size.

But still, de Moivre was answering the same question as Bernoulli.

That is: How likely am I to see this data, given a certain hypothesis? For instance, to take us back to the imagined bet a few paragraphs ago, how likely am I to see sixty or more heads out of one hundred coin flips? (The answer is: not very. Only about 2.8 percent of the time, in fact. That £50-to-£10 bet would be dreadful, and you shouldn't take it.)

What neither Bernoulli nor de Moivre were able to answer was what came to be known as *inverse probability*, although it's really the heart of what we want probability to do. What we want, or at least what science wants, out of statistics, is to answer: Given the results I've seen, what can I say about my hypothesis?

SIMPSON AND BAYES

Bernoulli and de Moivre and the rest were all perfectly well known to Bayes and his circle of well-to-do amateur mathematicians. "I think he and [Lord] Stanhope were studying the 1733 edition of de Moivre's *A Doctrine of Chances*," David Bellhouse, Bayes's biographer, told me. "I think it was studying that book which got Bayes interested in probability." That was probably around 1735, when Bayes was in his mid-thirties.

At the same time, another English Thomas, Thomas Simpson, was working on similar problems to de Moivre. Simpson was the son of a Leicestershire* weaver, and a weaver himself, who taught himself mathematics—apparently a relatively common occurrence; about half

---

*This is pronounced "Lester-sher," because England is a mysterious and contrary place.

the members of the Spitalfields Mathematical Society, which Simpson would later join, were weavers.[39] He seems to have had an interesting life: according to Stigler, at the age of nineteen he married a fifty-year-old widowed mother of two (although other biographies say that he married his landlady and they had two children of their own[40]), and the family had to flee from Nuneaton to Derby after "he or his assistant had frightened a girl by dressing up as a devil during an astrology session,"[41] apparently after a solar eclipse. By 1736 they were living in London.

The work of Simpson's that is most relevant to us here came in 1755. It was a treatise on measurement error in astronomy: If six astronomers all record the passage of a planet past a certain point, and all have slightly different results, what should we record as its true position?[42]

Simpson's answer was that we ought to use the mean of the observations, instead of (as some at the time suggested) the "Aristotelian mean," the largest result plus the smallest divided by two. He showed this, essentially, by demonstrating a special case of the law of large numbers.

I'm not going to go into the detail of it here, partly because it covers a lot of the same ground as standard deviation and so on, but it's notable for two key points. The first is that Simpson is explicitly talking about *inference*, rather than sampling. That is: "What can we say about the hypothesis, given the data?" rather than "How likely are we to see this data, given a hypothesis?" Simpson is trying to estimate the real position of the planet, not say how likely we are to see the errors we do, given a particular position. He could only do this by making some highly simplifying assumptions about the errors, but nonetheless, it's a genuine attempt to turn statistics into a useful inferential tool instead of a novelty (or a way of winning in casinos). He even gives what Stigler calls "the earliest statistical advice from mathematician to experimental scientist of which I am aware":[43] that we should use the mean of as many observations as possible.

The second key point is that one of the reviewers of Simpson's paper was the Reverend Thomas Bayes of Tunbridge Wells. "By then he was fairly mature in his use of probability," says Bellhouse, "and he gave some very insightful comments." The key one is what we would now call *measurement error*. "Bayes's major comment was to the effect that yeah, the math is right, but what if the measuring instrument is biased?" says Bellhouse. "Then the mean won't help you."[44]

In Bayes's words, from a letter he wrote to the physicist John Canton (another weaver's son):

Now that the errors arising from the imperfection of the instruments & the organs of sense shou'd be reduced to nothing or next to nothing only by multiplying the number of observations seems to me extremely incredible. On the contrary the more observations you make with an imperfect instrument the more certain it seems to be that the error in your conclusion will be proportional to the imperfection of the instrument made use of. For were it otherwise there would be little or no advantage in making your observations with a very accurate instrument rather than with a more ordinary one, in those cases where the observation cou'd be very often repeated: & yet this I think is what no one will pretend to say.[45]

Say you're trying to time how long it takes someone to run a mile, but the watches you use are all slightly fast so that the second hand goes around every fifty-nine seconds instead of every sixty; then taking the mean won't help, no matter how many watches you use—you'll just get ever more confident about your wrong answer.

Simpson appears to have taken this point, and included in a later version of his work a line that said his thinking only worked if "there is nothing in the construction, or position of the instrument whereby the errors are constantly made to tend the same way, but that the respective chances for their happening in excess, or in defect, are either accurately, or nearly, the same."[46]

We know, then, that Bayes was thinking about probability, and specifically *inferential* or *inverse* probability—remember, that's "How likely is it that a hypothesis is true, given this data?"—as opposed to *sampling* probability—which asks, "How likely am I to see this data, given this hypothesis?"—at least by 1755, and if Bellhouse is right, he'd been interested in the subject since reading de Moivre's book in the mid-1730s.

BAYES'S NOT-IN-FACT-A-BILLIARD-TABLE

One of Bayes's great contributions to probability theory was not mathematical, but philosophical. So far, we've been talking about probability as though it's all a real thing, out there in the world. The probability of a coin turning up heads *is* 0.5. The probability of seeing sixty or more heads in one hundred coin flips *is* about 2.8 percent. We say these things as though they're facts about the world. Bayes turned that around.

For Bayes—to quote Professor Sir David Spiegelhalter, the former president of the Royal Statistical Society, the former Winton Professor of the Public Understanding of Risk at Cambridge University, and surely the owner of the single most authoritative-sounding set of titles in all of statistical science—probability "is an expression of our lack of knowledge about the world."[47]

That is, for Bayes, probability is *subjective*. It's a statement about our ignorance and our best guesses of the truth. It's not a property of the world around us, but of our understanding of the world. If you flip a coin, and hide the result from me, and say, "What are the chances that the coin landed heads?" I might answer "fifty-fifty" if I trusted that you were doing it honestly. But if I knew you were a stage magician, or the owner of the world's largest collection of double-headed coins, I might make a different estimate.

What Bayes showed in his paper "An Essay towards Solving a Problem in the Doctrine of Chances" was that in order to make inferential probability work—which means, remember, asking, "What are the chances that my hypothesis is true, given the data?" rather than "What are the chances that I would see this data, given my hypothesis?"—you must take into account *how likely you thought the hypothesis was in the first place*. You must take your subjective beliefs into account.

To make his point, Bayes used a metaphor of a table, upon which balls are rolled. (Note that this is *not* a billiard table. "Later writers . . . have promoted it to a billiard table," sniffs Stigler, "but the Reverend Bayes was neither so specific nor so frivolous."[48] Spiegelhalter does call it a billiard table, but adds, "Being a Presbyterian minister, Bayes just called it a table."[49]) The table is hidden from your view, and a white ball is rolled on it in such a way that its final position is entirely random: "There shall be the same probability that it rests upon any one equal part of the plane as another."[50]

When the white ball comes to a rest, it is removed, and a line is drawn across the table where it was. You are not told where the line is. Then a number of red balls are also rolled onto the table. All you are told is how many of the balls lie to the left of the line, and how many to the right. You have to estimate where the line is.

Imagine that five balls are thrown, and you're told that two of them landed to the left of the line and three of them landed to the right.

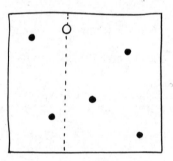

Where do you think the line ought to be? Bayes said that the most likely place is three-sevenths of the way up the table from the left.[51]

Intuitively you might think that it should be two-fifths. After all, you've just rolled five balls, and two of them ended up on one side and three on the other. But Bayes said that you must take into account the *prior probability*—your best guess of what the situation was, before you got any information.

But do you *have* a best guess? You don't know anything, do you? The line could be anywhere. But that in itself is a form of prior information: it is equally likely (from your subjective point of view) that the line is right up against the left-hand cushion, or right up against the right-hand cushion, or anywhere in between.

You could draw a graph of the distribution of probability—how likely the line is to be in a given place on the table—before you rolled another ball. It would look like this:

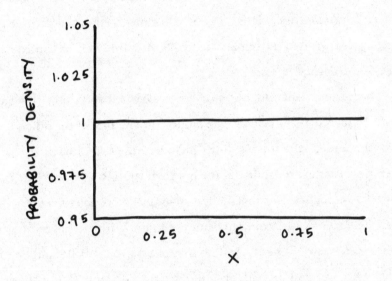

If you have absolutely zero idea where the line is, then the probability of the next ball landing to the left of it is 0.5—50 percent. After all, the line could be far to the right, so the ball would definitely land left; it could be far to the left, so the ball would definitely land right; it could be in the middle, so it would be fifty-fifty; or it could be anywhere else, with corresponding probabilities. The average position is exactly in the middle.

Essentially, Bayes's big insight was that you must add any new information you get to the information you already have. In this case, you don't have very much information. But it is something.

What that means is that instead of just saying, "The most likely position of the line is two-fifths of the way along the table," you have to take account of your prior. So Bayes said that the equation for working out the probability here is not "the number of red balls on the left divided by the total number of red balls"—⅖—but *the number of red balls on the left of the line PLUS ONE, divided by the total number of red balls PLUS TWO*. It is, says Spiegelhalter, "equivalent to having already thrown two 'imaginary' red balls, and one having landed each side of the dashed line."[52]

That might seem odd, but it makes sense when you think of what it would look like if all the balls had landed on one side or the other. If all five had landed left, and we didn't include those extra imaginary balls, then we'd say the probability of the next ball landing left was ⅗, or 1, or complete certainty. But that's silly—obviously you don't *know for sure* that the next ball isn't going to be to the right. With Bayes's extra balls, your estimate would be 6/7. And no matter how many balls land on one side, you never end up with absolute certainty—if a million balls land to the left, then your estimate of the next ball's probability of landing right would be 1/1,000,002. Each piece of new information pushes you closer to certainty, but you never quite get there.

What Bayes also did was talk about probability distributions. We know the most likely place where the line would be, but it's almost as likely to be a little bit either side of it, and a bit less likely to be farther away. And it's possible (but very unlikely) that it's way to the right, but three balls happened to squeeze into the space next to the cushion. You draw that probability distribution as a graph.

We saw what a uniform probability looks like a few paragraphs ago—a flat line across the graph. After the five balls have been rolled, you can redraw the graph, using some fairly complicated mathematics, and it would look like this:

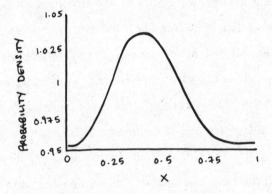

This is your *posterior* probability distribution—what your assessment of the likely position of the line looks like now that your prior has been updated with the new information.

But if you start going out to get more information, your posterior[*] becomes your new prior. If you were to roll another five balls, you'd go through the exact same process. And, most likely, your new distribution would be even narrower, and more precisely centred around the true value.

This exact system is what we do in all the examples we've looked

---

[*] Yes, yes, posterior. Get your chuckles out of the way now, we're not going to be able to avoid the word for the rest of the book.

at so far: when we use a medical screening test to look for cancer, or a COVID test to test for COVID, or evidence to convict a suspect. We're taking prior information (how common is the cancer?) and adding new information (a positive test with a certain level of sensitivity and specificity) and creating a new posterior distribution.

And, crucially, it's all subjective. That doesn't mean it's random, or that any prior probability is equally valid—if your prior is that you'll roll a six on a six-sided die roughly one-sixth of the time, and mine is that you'll do it five-sixths of the time, you'll probably be closer to the truth than I am, since most dice are fair. You can have better and worse reasons to believe something. But it is subjective. (Of course, if the die *is* fair and we roll it a few hundred times, and if I am a good Bayesian and update my beliefs appropriately in the light of new evidence, we'll see the six come up about one time in six, and I'll rapidly adjust my probability to something very close to yours.)

"An Essay towards Solving a Problem . . ." was most likely written in the period after Simpson's 1755 article. It appears to have sunk almost without trace—it was published after Bayes's death, but was apparently unknown to Pierre-Simon Laplace, the French mathematician who independently arrived at similar conclusions in 1774. Stigler argues that Bayes himself didn't think all that much of it—he wrote a will in 1760, four months before he died, suggesting that he knew he might not have long to live, and that he "would have had the opportunity to communicate his work to the Royal Society had he wished to do so,"[53] since he was by then a fellow. But he didn't. Instead, it ended up with his friend Richard Price, to whom Bayes left £100 in his will, along with his papers, including "An Essay towards Solving a Problem . . ." (Bayes didn't seem to know exactly where Price was: his will directed that the money and papers go to "Richard Price now I suppose preacher at Newington Green.")[54] But Price, it appears, had a better sense of the importance of the work.

PRICE, THE FIRST BAYESIAN, WANTS TO SAVE GOD FROM HUME

Richard Price (1723–1791) was another Nonconformist minister, who (as Bayes rightly guessed) was working at a chapel in Newington Green, North East London. The chapel is now rather famous—it is the oldest Nonconformist church in London still operating, and counted among its congregants Mary Wollstonecraft, author of *A Vindication of the Rights of Woman* and the mother of Mary Shelley, the author of *Frankenstein*.

At the time, Price was a much more well-known man than his older friend Bayes. He was well connected with radical thinkers: notably, he was friends with several of the Founding Fathers of the American Revolution. He exchanged letters with Thomas Jefferson[55] and Benjamin Franklin,[56] both of whom visited him in Newington Green, as did John Adams, the second president of the United States. Franklin in particular appears to have been a close friend. Price was a famous supporter of the revolution: his pamphlet *Observations on the Nature of Civil Liberty, the Principles of Government, and the Justice and Policy of the War with America* was published in February 1776, months before the Declaration of Independence; it sold out in three days and had been reprinted eleven times by May. According to a biography of Price, "the encouragement derived from this book had no inconsiderable share in determining the Americans to declare their independence."[57]

Price was also friends with the philosophers David Hume (of whom more later) and Adam Smith, and with William Pitt the Elder, the politician. In all he seems to have been a pretty cool and impressive guy with cool and impressive friends, famous across England and America, and it is strange how little known his name is now.

Price is important to this story as the man who brought Bayes's paper to wider attention: he showed the paper to the physicist John Canton in 1761, after Bayes's death, and had it published in the *Philosophical Transactions of the Royal Society* two years later.

Part of the reason it took so long for him to publish it was that Price didn't just go over it for typos and misplaced commas; he was more than, in the words of the historian of statistics Stephen Stigler, "a loyal secretary." Price had visions of his own for the work: while Bayes wrote the first half of the paper, the second half, containing all the possible practical applications of the theorem, was all Price.[58] Bayes had no interest in applied statistics: his work, in this paper and all others, was "all theory with not a hint of application."[59] But Price was—to quote Stigler again—"the first Bayesian."[60]

Price imagined a person "just brought forth into this world," seeing the sun for the first time: "after losing it the first night he would be entirely ignorant whether he should ever see it again," wrote Price in his appendix to Bayes's work. Then the next morning, it rises again, and again. After n mornings, how confident should you be that the sun will rise again on morning $n + 1$?

He argues that you can apply Bayes' theorem to the problem. Having seen the sun once, you have no idea if it's a one-off or a repeated event: its appearance simply shows that it is possible, but it could happen on one morning in every quadrillion, or every morning, or at any other rate. You should, as we would call it nowadays, have a uniform probability distribution over all the possibilities.

Once you see it return once, Price reasoned, you should be sure that "there was an odds of 3 to 1 for some probability of this." It's just like Bayes's ball table, except instead of working out the odds of the ball landing to one side or other of the line, you're working out the odds of the sun coming up. Once you see it return a million times, "there would be odds of the millionth power of 2, to one, that it was likely that it would return again at the end of the usual interval." But no amount of evidence "would be sufficient to produce absolute, or physical certainty."[61]

Which is all good fun, and in keeping with the steady progression we've been seeing, of Bernoulli trying to work out what degree of "moral certainty" you could achieve with a given sample size, and of Simpson and de Moivre trying to turn the idea of *sampling probability*, the chance of seeing the data we've seen if we assume a hypothesis is true, into *inferential probability*, the chance that a hypothesis is true given the data we've seen.

But what's interesting is what motivated Price to do that. At the time, there was a divide among the Nonconformist ministries, between those who thought mathematics would lead to godlessness and those who thought it would help us understand God's universe. Price was of the latter persuasion, and so, Stigler and Bellhouse believe, he wanted to use Bayes' theorem to save God from David Hume.

Hume, in his 1748 essay "Of Miracles," argued that no amount of testimony should ever convince someone that a miracle, a violation of natural law, took place: he never actually said "extraordinary claims require extraordinary evidence," but that's the gist of it. "[No] testimony is sufficient to establish a miracle," wrote Hume, "unless the testimony be of such a kind, that its falsehood would be more miraculous, than the fact, which it endeavours to establish." If someone were to say that he had seen the dead restored to life, Hume continues, "I immediately consider with myself, whether it be more probable, that this person should either deceive or be deceived, or that the fact, which he relates, should really have happened." [62]

This was pretty shocking stuff to a Christian nation, who firmly believed in at least one person coming back from the dead in the New Testament, and Hume's essay was met with a hostile reaction. But the point is one of probability: we all have a lifetime's experience of the laws of nature not being broken, and we also have a lifetime's experience of people saying things that are not true. If someone says, "I saw a dead man come back to life," most of us would consider it more likely that that someone is wrong, or lying, than that they actually saw a dead man come back to life. So, says Hume, we should ignore that testimony as irrelevant.

But Price, newly armed with Bayes' theorem, wanted to say that rare events *do* happen, and that even if you've seen the sun rise or the tide come in a million times, you can never be *physically certain*, in his phrase, that it'll do so the next time. In his appendix to Bayes's work he wrote a long argument about a die with an unknown but very large number of sides, which comes up showing a certain face a certain number of times over a million throws, and what you could conclude about that. He used the exact same numbers in a later essay, explicitly attacking Hume's "Of Miracles."[63]

In the later, Bayes-derived work, Price imagined this die, with a huge but unknown number of faces, perhaps a million or more. Some faces are marked, let's say with an X, and some aren't, but you have no idea what the ratio is and you're trying to work it out. You roll the die a million times. On all 1 million rolls, you see an X.

This is exactly analogous to Bayes's not-in-fact-a-billiard-table we discussed in the last section. Instead of "Has the ball fallen to the left or the right of the line?" it's "Has the die shown an X or not?" but you can do the same math precisely. And if you roll X 1 million times in a row—and, importantly, you went into the situation with no knowledge of the likelihood of X versus not-X—then your best estimate of the probability of the next roll coming up not-X is 1/1,000,002. And the *distribution* of the probability—the curve on the graph—is centered around that figure. Specifically, Price calculated, there is a 50 percent chance that the probability of not seeing an X lies between 1 in 600,000 and 1 in 3 million.

Price then goes on: Say you're not talking about a die. Say you're talking about watching the tide come in, twice every day. You've seen it a million times (you're fourteen hundred years old). There is still a small, but real, possibility that on the million-and-first time, it just won't. Rare events happen sometimes, and no amount of seeing them *not* happen will ever completely rule them out. Similarly—Price would say—you might have seen people fail to rise from the dead a large number of times, but you can't ever say with certainty that it never happens.

Hume saw Price's work. In fact there was a rather lovely exchange between the two men, which I will tell you about just because it's so nice to see two people who disagree so profoundly on something so important—God versus no God—behaving so civilly.

Price included some mildly rude lines in his essay responding to "Of Miracles," such as one suggesting that someone putting forward arguments such as Hume's "would deserve more to be laughed at than argued with." Then the two men met, and Hume, by all accounts a very affable and reasonable man, left Price both charmed and ashamed: in a second edition, he removed every disobliging comment ("it is indeed nothing but a poor though specious sophism" replaced with "I cannot hesitate in asserting it is founded on false principles," for instance) and added a rather apologetic introduction saying that one shouldn't accuse one's opponent of bad faith or disingenuousness.[64] And Hume, after Price's apology, sent him a sweet letter saying that he had nothing to apologize for, that he was "a true Philosopher," that he had treated Hume "with unusual Civility . . . as a man mistaken, but capable of Reason and conviction,"[65] and that Price's arguments were "new and plausible and ingenious," although Hume never revisited the essay.

Price had actually gone further than this in the foreword to Bayes's paper. He didn't make any claims about miracles, but he suggested that Bayes' theorem could show that the world progressed according to fixed laws, and thus could "confirm the argument taken from final causes for the existence of the Deity." I am not sure how many modern statisticians would agree with that, but right from the start, Bayes' theorem had some lofty applications.

FROM BAYES TO GALTON

Bernoulli, de Moivre, and Simpson had, between them, shown that if you take lots of measurements of a thing, and (as Bayes pointed out) as long as the errors in those measurements are random rather than systematic, then those measurements will tend to center around the true value.

Bayes had shown that if you take into account a prior estimate of what that true value is most likely to be, you can use those measurements to make inferences—to make statements about what is likely in the world.

In the years after Bayes's death, the great French mathematician and physicist Pierre-Simon Laplace independently arrived at the same conclusions as Bayes, and gave a rather more detailed account of it. Richard Price visited Paris in 1781 and discussed Bayes' rule with Nicolas de Caritat, Marquis of Condorcet, who was a mentor of Laplace's. Condorcet, and then Laplace himself, went on to acknowledge that Bayes got there first, hence "Bayes' theorem" rather than "Laplace's theorem,"[66] even though Laplace's treatment of the problem was probably the more impressive.

Probability theory had grown out of games of chance, and out of physics—the main use had been in astronomy, trying to use the average of several observations to minimize the overall error. But its use in the social sciences was obvious—Jacob Bernoulli had talked about the actuary's problem of working out how likely someone was to live for another ten years by looking at other people of similar age and status:

[If] from among the observed three hundred men of the same age and complexion as Titius now is and has, two hundred died after ten years with the others still remaining alive, we may conclude with sufficient confidence that Titius also has twice as many cases for paying his debt to nature during the next ten years than for crossing this border.[67]

Laplace, sixty or so years later, looked at birth rates in Paris and found that there was a small but real bias toward boy babies—251,527 boys were born in the city between 1745 and 1770, compared to 241,945 girls (a roughly 51:49 ratio), and declared that there was only about a 1 in 10 ^ 42 chance that you'd see a result that extreme if each birth was equally likely to be a boy or a girl.[68] He also noted that there was an even more extreme bias among London births.

But it was Adolphe Quetelet (1796–1874), a Belgian mathematical prodigy, who really pushed probability and statistics into the social sciences. He worked as an astronomer and meteorologist at the Brussels Royal Observatory, but his outside interests were in statistics, and by the age of twenty-six he had become a senior figure in the national statistical office, analyzing population data and organizing a census. He also came close to the idea of randomly sampling a population to get a sense of the whole, as in an opinion poll, and using that *instead* of a census, following a model of Laplace's. But he was talked out of it by the Baron de Keverberg, who argued that you could never be sure that your sample was truly representative of the population at large because there was too much variation in the subpopulations.

Quetelet's main contribution was that of the "average man." He gathered data about people—not just men—along various axes: physical attributes such as height, weight, strength; moral and psychological ones such as drinking, crime, insanity. He wanted to find the average along each of these axes, as the fundamental units for what he would call "social physics." You could analyze society along these different axes: How do education levels interact with how likely it is that someone is convicted of a crime? How about literacy? How about age?

What Quetelet noticed was that many of these measurements were normally distributed—he looked at the chest circumference of Scottish soldiers, for instance, and found the normal curve, similar to how, if you measured the same soldier's chest several times, your results would be distributed (by measurement error) around a mean. He suggested that it meant that things like height, weight, strength, and even behavioral characteristics, like suicide, were the product of many small influences, and it was rare for all those influences to point in one direction or another; usually some would point one way, some another, and so people's heights, weights, and drinking habits would tend to cluster around the population mean, in a normal distribution. "It would be as though each person's height had been determined by drawing some large number of pebbles from the same urn with a given fixed urn-ratio," writes Clayton, "with each white pebble making them taller and each black pebble making them smaller." [69]

Quetelet wanted to find laws of human society analogous to the laws of physics, but he also started to think that the *average man* was in some way the ideal person: "a standard of beauty at which nature aims," in Stigler's phrase, [70] although others suggested that the average person would be mediocre or even somehow monstrous. Quetelet made various mistakes—he hadn't realized that there are other ways for a quantity to be normally distributed, for instance, and also went around gleefully applying the normal curve to everything he saw. A later statistician would diagnose the condition of "Quetelismus,"[71] of seeing the normal distribution everywhere you look.

But Quetelet's obsession with the normal distribution caught on somewhat. And his work led to the idea that you could make probabilistic predictions about individual behavior and life outcomes by looking at the wider population: he saw, for instance, in a famous work on jury trials, that there were differences in the likelihood of being convicted depending on whether the defendant was male or female, over thirty or under thirty, well educated or not, literate or not. (If you were hoping to be acquitted, you would prefer to be a well-educated woman over thirty, by the way, or at least you would in early nineteenth-century France.)

This was hugely controversial: it seemed to be in conflict with the idea of free will, that our behaviors and choices were the product of our attributes. It also set the scene for what would later be called "scientific racism": one follower of Quetelet's, Alphonse Bertillon, found that among the young male population of a town called Doub, there seemed to be *two* peaks to the curve, as there would be if you measured the heights of men and women; there were *two* "average men" in Doub. He suggested that this was because Doub's inhabitants were of two races, the Celts and the Burgundians.[72] This turned out to be a mistake: it was shown some years later that Bertillon had screwed up as he converted from inches to centimeters and had made the data look as though it were saying something it wasn't. But it paved the way for others to follow.

And—for Clayton, at least—the controversy over measuring humans scared statisticians away from the Bayes/Laplace model of acknowledging the subjective nature of probability theory and made them want to hide behind apparently objective, apparently solid statistics.

It was shortly after this that Francis Galton entered the scene.

GALTON/PEARSON/FISHER
AND THE RISE OF FREQUENTISM

Despite Bayes's and Laplace's work, statisticians and scientists don't, on the whole, use Bayes' theorem in their everyday work. Instead, most are what are called *frequentists*.

Frequentist statistics do the opposite of what we've been talking about. Where Bayes' theorem takes you from data to hypothesis—How likely is the hypothesis to be true, given the data I've seen?—frequentist statistics take you from hypothesis to data: How likely am I to see this data, assuming a given hypothesis is true?

That, of course, is what Bernoulli had been doing more than a century before, and what others had been trying to get beyond. So why did it revert?

Before I go any further: in case it isn't clear already, this is all *phenomenally* controversial stuff. I have written about scientific controversies for quite a few years now, and they really do get pretty tasty, but the Bayesian-frequentist "stats wars" are probably the most ill-tempered of the lot. So what follows will annoy many people even if I get the basic shape of it right.

But my basic understanding goes like this: priors are a problem.

They're a problem for technical and pragmatic reasons: How do you choose them? On Bayes's imaginary not-in-fact-a-billiard-table, he assumed that it was equally likely that the white ball could be anywhere on the table. That's called a *uniform prior*. That's defensible—you can imagine that if you throw the ball hard enough it'll be essentially random where it lands. But what about situations where you're completely ignorant and don't have good reasons to assume any prior?

A more technical objection was that of the mathematician and logician George Boole, who pointed out that there are different kinds of ignorance. A simplified example taken from Clayton:[73] Say that you have an urn with two balls in it. You know the balls are either black or white. Do you assume that two black balls, one black ball, and zero black balls are all equally likely outcomes? Or do you assume that *each ball* is equally likely to be black or white?

This really matters. In the first example, your prior probabilities are one-third for each outcome. In the second, you have a binomial distribution: there's only one way to get two black balls or zero black balls, but two ways to get one of each. So your prior probabilities are one-quarter for two blacks, one-half for one of each, one-quarter for two whites.

Your two different kinds of ignorance are completely at odds with each other. If you imagine your urn contains not 4 but 10,000 balls, under the first kind of ignorance, your urn is equally likely to contain 1 black and 9,999 whites as it is 5,000 of each. But under the second kind of ignorance, that would be like saying you're just as likely to see 9,999 heads out of 10,000 coin flips as you are 5,000, which of course is not the case. Under that second kind of ignorance, you know you're *far* more likely to see a roughly 50–50 split than a 90–10 or 100–0 split in a large urn with hundreds or thousands of balls, even though you're supposed to be ignorant.

So which prior do we assume? Do we think the color of the balls is independent or correlated? You may say that you assume perfect ignorance, but there are different kinds of "ignorance," and you have to pick one.

But the underlying problem of Bayesian priors is a philosophical one: they're *subjective*. As we said earlier, they're a statement not about the world but about our own knowledge and ignorance.[*] And that's . . . uncomfortable. The promise of science and numbers—still, today, but I think even more so in the eighteenth and nineteenth centuries, when people like Quetelet could use terms like "social physics" without actual physicists sniggering behind their hands—was one of objectivity. We should all be able to look at the outcome of some experiment, or some observational study of the chest measurements of Scottish soldiers, or whatever, and agree on what it says.

What the Bayesian model seems to say is that whether something is true or not depends on *how strongly I believed it before*. So if we carry out a study on homeopathy or the Higgs boson and find some positive result, then you might think that result very likely to be real, and I might not, and we might both be correct to do so—if our prior probabilities were sufficiently different.

---

[*] Some Bayesians—Harold Jeffreys and, notably, E. T. Jaynes—do talk about "objective Bayesianism," trying to base priors on logical principles. "I don't think [Jaynes] succeeded," Kevin McConway of the Open University told me, "but he did have a good try."

There's something soft and squishy about the idea that probability is ultimately subjective and personal, rather than something real, out in the world. If I say, "There's a 50 percent chance this coin will come up heads," it feels like I should be making some statement about the coin, not about my own *beliefs* about the coin.

Of course—and we'll come back to this—"subjective" doesn't mean "random," or "baseless." Say I have two beliefs: one, that a fair coin tossed fairly will land heads 50 percent of the time, and two, that there is a 90 percent chance that I will be abducted by aliens tomorrow. They're both subjective statements about my internal beliefs, but most people would agree that the first belief is reasonable, while the second one is not. But nonetheless, the idea that probability is *in your head*— that when we say, "There's about a one-in-six chance this die will come up six," we're making a statement about our beliefs rather than about dice—was not, and is not, universally accepted by statistical thinkers.

It was this distaste for subjectivity that seems to have driven the rise of frequentism.

ARE FREQUENTISTS RACIST?

It's here that we get into some serious controversy. First, we need to acknowledge that, by twenty-first-century standards (and arguably by the standards of their own time), some of the people involved in what you might call the golden age of statistical thought held pretty appalling views. The question is whether we can separate those views from the statistical theories these men came up with.

Francis Galton (1822–1911), for instance, was in many ways an extraordinary man: "perhaps the last of the gentleman scientists."[74] The cousin of Charles Darwin and a qualified doctor who inherited a fortune, he quit medicine and went off to do whatever took his fancy. He explored Africa and was awarded a medal for doing so by the Royal Geographical Society; he got meteorological stations to fill out surveys about the weather and, using the data, became the first to notice "anticyclones," the whirlpools in the air that we are familiar with from satellite images; and, crucially, he pushed forward the use of statistics in studying humans and, specifically, how things like talent are passed on through families.

Galton spent most of his career at University College London, where he made various huge breakthroughs. For example: there was a confusing problem with the "normal distribution." Imagine that you're looking at the size of grapes. You might expect them to be normally distributed: the average-size ones being the most common, while very large ones and very small ones would be rarer. But imagine now that the grapes are grown in three areas on the same hill: the north face, the east and west faces, and the south face. The north face gets the least sun, so its grapes are smallest (on average). The east and west faces get more, so their grapes are larger. The south face gets the most, so its grapes are largest.

Each of *those* sets of grapes would, you'd expect, be normally distributed. But then . . . does that mean that the overall set of grapes *isn't* normally distributed? Should we see three peaks on the graph? And what if there were more than three groups, or if there was a sliding scale of inputs? Or if the amount of rainfall differed, as well as the sunshine? How come, despite all these different inputs, we see a normal distribution overall?

Galton was not an especially talented mathematician, so to solve the problem he used a clever alternative, a thing called a quincunx, which looks a bit like a fairground game. It's a big board with an array of pins on it and, at the bottom, a series of compartments. At the top it has a funnel to pour little ball bearings into. The ball bearings fall down the board, bouncing (in theory) randomly left or right off the pins, and landing at the bottom. The balls (as you'd expect) tend to land in an approximation of the normal curve: the random bouncing tends to even itself out, in a way that de Moivre would recognize, but sometimes you get a lot of rightward bounces.

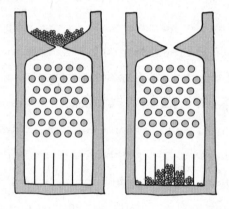

Galton's insight was to add a second layer of compartments above the first, which traps the balls higher up the board and that could be released individually.[75] The balls would lie in a normal distribution in the higher level; but when you released one compartment, the balls in that compartment would form a normally distributed curve below it. What Galton showed was that if you released all the compartments, even though each one formed a normal curve on its own, they added up to a bigger normal curve.

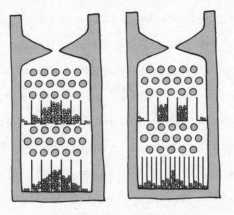

What this showed, he said, was that in situations like the grapes, lots of smaller normal distributions could add up to one big one, as long as the middle-size distributions were more common (as in our example, where there are two medium-sun faces, the east and west slopes, and only one each of the extremes, the north and the south). This allowed Galton and later statisticians to think about lots of different populations forming part of a larger one.

Galton was also the first to explain what we now know as *regression to the mean*, or as he called it, *regression to mediocrity*.[76] Looking at sweet peas, but thinking about humans, he noticed that the offspring of very tall parents tend not to be as tall as their parents, compared to an average of the two parents' height; and the offspring of very short parents tend to be rather taller. This was confusing, because you might expect the offspring's heights to be normally distributed around their parents. Galton showed that this was a general finding: any two variables that correlated somewhat but imperfectly (so, for example, parents' height and their offspring's height, or an individual's height and his or her weight, or a country's population and its GDP) would demonstrate the same phenomenon. If you find an outlier on one variable—someone very heavy, say, or a country with a particularly enormous GDP—it's likely that they'll be less extreme on the other variable, just because extreme values are unlikely.

Galton was also very interested in the inheritance of talent. He wrote a book called *Hereditary Genius,* looking at how brilliant thinkers often clustered in families (his own family, with Erasmus Darwin and Charles Darwin as close relatives, may have been an inspiration). He coined the phrase "nature and nurture" to refer to the twin inputs of heredity (what we'd now call genetics) and environment. But what he really wanted to do was to create a science of human breeding—*eugenics,* another phrase which he came up with.

I want to be careful here, because there's a tendency among some commentators to link *all* research into human intelligence and its heritable nature with eugenics or "scientific racism." We really can measure human intelligence pretty well, with the imperfect but generally useful metric of IQ, and it really is heritable—clever studies designed to separate the input of genes and environment show that about half of the variance in IQ is caused by the genes we inherit from our parents.[77] It's a well-established, much-repeated finding, and intelligence research is good, important science.

But Galton didn't just want to observe and document facts about how intelligence is distributed. He wanted to breed humans. "If a twentieth part of the costs and pains were spent in measures for the improvement of the human race that is spent on the improvement of the breed of horses and cattle, what a galaxy of geniuses might we not create!" he wrote.[78] "We might introduce prophets and high priests of civilisation into the world." He was, therefore, in favor of encouraging breeding among highly successful families, and discouraging it among less successful ones.

And he was extremely racist.[79] He wrote a letter to the London *Times* calling African people "inferior" and "lazy, palavering savages," saying that "the Arab" is "little more than an eater up of other men's produce; he is a destroyer rather than a creator," and that East Africa should be handed over to the Chinese because, while they are given to "lying and servility," that is the product of their education, and "Chinamen" are by nature "industrious [and] order loving."[80] (Anglo-Saxons were, for Galton, the best extant race, although the best of all time were the ancient Athenians: "The average ability of the Athenian race is, on the lowest possible estimate, very nearly two grades [standard deviations] higher than our own"—a comparable difference, he said, to that between Anglo-Saxons and Africans.) He was obsessed with cataloguing and comparing the races, with the tools of science that he himself helped to create.

Galton's work inspired a later generation of statisticians—notably Karl Pearson (1857–1936), and after him Ronald Fisher (1890–1962). Like Galton, Fisher and Pearson were brilliant, and like Galton they were, certainly by the standards of our day and arguably by the standards of their own, unpleasantly race-obsessed. Also, they *hated* each other.

Pearson was a polymath, a historian, philosopher, physicist, lawyer, and politician before he was a mathematician. In 1885 he became a professor of applied mathematics at UCL, following in Galton's footsteps. Galton, upon his death, left money to UCL to found a chair of eugenics, and Pearson was the first appointee.

Along with Galton and a third man, Raphael Weldon, Pearson founded a journal of statistics, *Biometrika*. He came up with the "chi-square test," which allowed mathematicians to check whether a sample of data really was normally distributed, or whether it best fit some other curve. He also was the first to coin the term "standard deviation."

Fisher was younger; he was appointed the University College of London professor of eugenics after Pearson retired—or rather, the post was split in two, with Pearson's son Egon taking the other half. (If your eyebrows went up a little at the idea that UCL had a "professor of eugenics," then don't worry, we'll come back to that.) Fisher is a titan of statistical theory; "the dominant figure" of twentieth-century statistics,[81] according to the American statistician Bradley Efron. The list of modern statistical tools he invented or extended is remarkable. He was responsible for the various models used in "analysis of variance" (ANOVA), for the concept of "statistical significance," for the "maximum likelihood estimation" (MLE) method for establishing which hypothesis about the distribution of data would best explain some given data, and for a host of other things. Fisher was also a pioneering geneticist: when statisticians speak to life scientists, the statisticians are surprised to hear Fisher described as a great geneticist, when they think of him as a great statistician, and vice versa.

Both men tried to move statistics away from where Laplace and Bayes had left it, relying on subjective priors. Ironically, the reason they fell out was over Bayes. Specifically, it was about the maximum likelihood method. "Likelihood," in Fisher's new jargon, was essentially a way of saying how likely one particular hypothesis was, given some data, compared to another. For example, imagine you flip eight heads on ten coins. That's pretty unlikely on a fair coin: it would only happen about one time in twenty. But if you had a dodgy coin somehow, one that came up heads 80 percent of the time, then you'd expect to see exactly eight heads about one time in three. You're about seven times as likely to see this data under the hypothesis "this coin is biased and comes up heads eight times out of ten" than under the hypothesis "this coin is fair." So the likelihood ratio between these two hypotheses is about seven.

Fisher published this in a paper in Pearson's journal *Biometrika*.[82] But Pearson read it, and thought Fisher was sneaking Bayes in the back door. You can see why—it does look a bit like Fisher's MLE is inverse probability. It's sort of like it's saying, "This hypothesis is more likely than this one." But actually it's not: you might still be more likely to see a fair coin throwing eight heads than a dodgy one, *if there are lots more fair coins than dodgy ones around.* All the MLE does is let you compare how likely you are to see these results if we assume one hypothesis or another, but it doesn't, in its own right, tell you which hypothesis is the more likely.

But Pearson thought it did, and with some other authors he added an appendix to Fisher's paper essentially saying it was Bayesian—that it assumed you were treating the prior probability of each hypothesis as equal—and showed that (under those assumptions) it was incorrect. This completely blew their friendship apart (Fisher really didn't like being called Bayesian), and until Fisher's death he would continue feuding with (the by then long dead) Pearson; Egon Pearson, his son; and Jerzy Neyman, Pearson's successor.

Like Galton, both Pearson and Fisher had what we would now consider pretty unpleasant views. Specifically, they both were big fans of eugenics.

Again, I want to be careful here. Many at the time who were very much on the progressive, liberal end of society were also pro-eugenics. Marie Stopes, the campaigner for birth control, abortion, and women's rights, was also a major supporter of eugenics. John Maynard Keynes, the great economist and liberal (and my great-great-uncle, so I ought to declare an interest here when I try to downplay the awfulness of it all), was another. Sidney and Beatrice Webb, George Bernard Shaw, Bertrand Russell, all heroes of the socialist and liberal movements, were in favor of selective breeding of humanity in order to create a better, more perfect society. The term wasn't, as it is now, associated so heavily with the right. (In fact, when I was writing stories in the late 2000s and early 2010s about things like embryo screening for disability, in vitro fertilization, and mitochondrial donation, misleadingly named "three-person babies," it was mainly the religious right who criticized them as "eugenics.")

When Fisher wrote in Galton's journal *Eugenics Review*,[83] for instance, that "the nations whose institutions, laws, traditions and ideals, tend to the production of better and fitter men and women" will "supplant" those "whose organisation tends to breed decadence," it was probably not as shocking as it would be today. Or when Pearson—a socialist and liberal with for-his-time progressive views on women's emancipation—wrote of the "alien Jewish population" being "somewhat inferior physically and mentally to the native population."[84] Anti-Semitism was rife among the British liberal intelligentsia (and the rest of Britain), even as they admired Jewish thinkers like Theodore von Kármán or Baruch Spinoza. People in the past had opinions that we have rightly discarded, on the whole, today—even Darwin, a very liberal man by the standards of his day, had views that we would see as enormously racist now.

What's a bit more interesting, though, is whether Galton, Pearson, and Fisher's views on eugenics affected their views on science. Clayton argues forcefully that they did. "As far as the history of statistics and eugenics go," he told me, "they're intertwined. It's a necessary part of the story." At heart, he said, Fisher and to some extent Pearson hated the idea of Bayesianism because they wanted a veneer of objectivity for their eugenic views. If it was *science* that some races were inferior and others superior, if it was *objective truth* that we ought to discourage breeding among the poor, then we couldn't argue with it. Bayesianism and its inherent subjectivity, its squishy "What do I think?" nature, undermined that, Clayton said. "They sought out a kind of scientific authority," he told me, "because they knew they'd encounter resistance for this quite radical upheaval. They wanted that backed by the most unassailable authority possible."

And he backs this up by quoting Pearson himself. "We believe there is no institution more capable of impartial statistical inquiry than the Galton Laboratory," Pearson wrote in the foreword to his paper about the inferiority of Jewish children. "We firmly believe that we have no political, no religious and no social prejudices. . . . We rejoice in numbers and figures for their own sake and, subject to human fallibility, collect our data—as all scientists must do—to find out the truth that is in them."[85]

It would be possible to write whole books about the eugenics movement and how it intertwined with early science. Many have—Clayton's own book, *Bernoulli's Fallacy*, goes into it at length, and Adam Rutherford's *Control* describes how much of modern culture has its roots in eugenic ideas, largely starting with Galton.

But Clayton's claim is that the whole history of frequentist statistics went the way it did because of a need to drive eugenics. I don't think that's fair. Clayton is admirably honest about his own motivations—he admits there's a war between frequentists and Bayesians, and says in his book, "Consider this [book] a piece of wartime propaganda, designed to be printed on leaflets and dropped from planes over enemy territory to win the hearts and minds of those who may as yet be uncommitted to one side or the other. My goal with this book is not to broker a peace treaty; my goal is to win the war." It would certainly help win the war if it turned out that frequentists were racist.

David Bellhouse, Bayes's biographer and a statistician himself, is skeptical. "I wouldn't buy that line at all," he said. "That doesn't make eugenics OK, and it doesn't help disentangle the history of early twentieth-century science from its sordid links to white-supremacist movements; some of the Nazi race ideology can be traced back without too much difficulty to Galton, for instance. But that's for other books to talk about; the question I'm interested in is 'Which is correct?' or, perhaps more accurately, 'Which is more useful?' rather than 'Which had the more unpleasant adherents?'"

THE FALL OF BAYESIANISM

Galton, Fisher, and Pearson weren't the only people criticizing the Bayesian approach. The problem that people had, on the whole, was the idea that *if you don't know which outcome is the most likely, then you should treat them as equally likely.* John Stuart Mill, who was briefly a critic of the Laplace/Bayes model, wrote in 1843: "To pronounce two events equally probable, it is not enough that we should know that one or the other must happen, and should have no ground for conjecturing which. Experience must have shown that the two events are of equally frequent occurrence."[86]

For Mill, the idea that probability is just an expression of our ignorance was silly. Probability, he thought, expressed something real about the world: the frequency with which events occur. "Why, in tossing up a halfpenny, do we reckon it equally probable that we shall throw cross or pile?" he wrote. "Because experience has shown that in any great number of throws, cross and pile are thrown about equally often; and that the more throws we make the more nearly the equality is perfect." The Laplace/Bayes inverse probability, he said, implied that by doing clever things with numbers, "our ignorance can be coined into science."

This is as neat a description of the Bayesian-frequentist disagreement as you could ask for, I think. Bayesianism treats probability as subjective: a statement about our ignorance of the world. Frequentists treat it as objective: a statement about how often some outcome will happen, if you do it a huge number of times.

As we've discussed, there were specific criticisms. Boole noted the problem that different kinds of ignorance lead to different priors: Are we ignorant of the overall distribution of the balls in the bag, or of the color of each individual ball? A related problem was named after the French mathematician Joseph Bertrand[87] (although I've taken this adapted version from Clayton). Imagine someone draws a square and asks you to guess its size. It could be anything between zero and ten centimeters along each side.

If we assume a uniform prior—that is, that any length is equally likely—then we should also say that a 1 cm × 1 cm square is just as likely as a 9 cm × 9 cm square.

But on the other hand, surely we should also say we're uniformly ignorant of the *area* of the box. The largest area it could have is 100 square centimeters (10 cm × 10 cm). If we're assuming uniform ignorance, then a piece of paper less than 50 square centimeters in area should be just as likely as a piece of paper *more* than 50 square centimeters in area.

The trouble is that those two claims can't both be true. If we're uniformly ignorant of the length of the sides, then it's more than 70 percent likely that the square will have an area less than 50 square centimeters. (A 7 cm × 7 cm square would have an area of 49 square centimeters, because 7 × 7 = 49.) Meanwhile, if we're uniformly ignorant of the area, then it's 75 per cent likely that the sides will be at least 5 cm long (5 × 5 = 25). Again, as with Boole's criticism, there are different kinds of ignorance, and we are ignorant of which one to use. (Later Bayesian thinkers would introduce the idea of "higher-level priors," describing your ignorance of which prior to use.)

John Venn, the English philosopher and creator of the famous diagrams, who was a teacher of Fisher's at Cambridge, took Mill's idea that probability is about how often things actually happen in the real world, not about our estimates of how likely they are, and expanded on it. For him, when we say, "A fair coin will come up heads 50 percent of the time," what we mean is "If we flipped the coin an infinite number of times, it would come up heads in half of those flips." Of course, we can't *actually* flip a coin an infinite number of times. But, said Venn, we should *imagine doing so*.[88]

Fisher followed Venn and was explicit about it. "[When] we say that the probability of a five with a die is one-sixth, we must not be taken to mean that of any six throws with that die one and one only will necessarily be a five; or that of any six million throws, exactly one million will be fives; but that of a hypothetical population of an infinite number of throws, with the die in its original condition, exactly one-sixth will be fives."[89]

There's a fun little addendum to this. Fisher gave Venn and Boole, and the mathematician George Chrystal, credit for undermining Bayesianism. "The first serious criticism [of Bayesianism] was developed by Boole," he wrote. "In the latter half of the nineteenth century the theory of inverse probability was rejected more decisively by Venn and by Chrystal."[90]

But none of these men actually did that.[91] Boole did point out problems with the concept of a uniform prior, but didn't propose that the whole concept be abandoned, just that it presented a difficulty. He wrote, in a very Bayesian, subjective way, that "all the procedure of the theory of probabilities is founded on the mental construction of the problem from some hypothesis,"[92] and acknowledged that a principle of ignorance is a good starting point.

Venn, meanwhile, was also critical, but only of a subset of Bayesian thinking: the rule of succession, the idea that (as we talked about when we discussed Bayes's not-in-fact-a-billiard-table) the chance that an event that has happened n times in x attempts will happen the next time with probability $(n + 1)/(x + 2)$. Remember: if you roll the ball on the table five times, and it ends up to the left of the line twice, the chance that it will do so the sixth time is not ⅖ but 3/7. Fisher himself was highly critical of Venn's reasoning on this point, and the criticisms would have been equally damaging to Fisher's own work.

Chrystal, hilariously, had simply made a mistake. He had applied Bayes' theorem to another version of Bertrand's box paradox, and noted that it implied that the chance of drawing a white ball from a bag in certain circumstances was three to one in favor. But, he said, it was obvious that the *real* answer was fifty-fifty. So Bayes' theorem was wrong. But it wasn't: three to one was in fact the correct answer, and Chrystal had been led astray by his own intuitions.

Of the three thinkers Fisher called upon to support his attack on Bayesianism, none of them really did the job: it was Fisher himself (and Neyman, and to some extent Pearson) who did it. But nonetheless, Bayesianism was rendered unfashionable for a long time, and frequentist Fisherian/Pearsonian statistics are very much the standard among professional statisticians and scientists today. Fisher, who called Bayes' theorem a "staggering falsity,"[93] "perhaps the only mistake to which the mathematical world has so deeply committed itself,"[94] which "must be wholly rejected,"[95] had won.

STATISTICAL SIGNIFICANCE

I probably ought to tell you what frequentist statistics actually involves. There's lots to it, but at its heart it is the not-inverse probability: it is sampling probability, the probability that Bernoulli (or even Fermat and Pascal) would have recognized. It is *What is the chance of seeing this result, given some hypothesis?*

If you've ever read any stories about science in the media, you'll probably recognize the phrase "statistical significance." You may also have come across "p-values."

A p-value is the likelihood of seeing results at least as extreme as those you've seen, given the *null hypothesis*, which is to say, the hypothesis that whatever effect you're looking for *isn't real.*

Imagine you're looking at some data you've gathered. Let's say you've got data on people's IQ, and also their shoe size, and you want to see whether people with big feet tend to be cleverer. By definition, the average IQ is 100. You have fifty people with above-average-size feet—say, people with US size thirteens and over. You give all of them an IQ test, and you see that the average score is 103.

But, of course, it's a relatively small group of people, and (as Bernoulli noted) any idiot could tell you that small sample sizes are less reliable than larger ones, and fifty isn't the biggest sample size in the world. What can you say about your hypothesis, given your results? What frequentist statistics, of the Fisherian tradition, would do, is imagine that you knew that there was no effect—that people with big feet were no more likely to have high IQs than the rest of the population. That is your *null hypothesis.*

You then calculate how likely it would be to see data like those you're seeing, in precisely the Bernoulli fashion, under the null hypothesis, and you call that your p-value. If, say, you'd only see results at least as extreme as the ones you've seen one time in every ten, then your p-value is one over ten, or 0.1.

I've just stuck the "IQ and feet" example into an online calculator, made a few assumptions about the sample, and got a p-value of about 0.16. In essence, that means that if the greater-footed citizen was no more likely to have a high IQ than anybody else, and you picked groups of fifty of them at random from the population, then about one time in every six, you'd find that your group was at least this different from the population average, whether higher or lower.

But what does that mean? Can I or can I not say that big-footed people are cleverer?

What Fisher did was suggest that we should choose some arbitrary level at which we say, "OK, it's pretty unlikely that we'd see results this extreme given the null hypothesis, so I'm going to behave as though it's a real effect."

Fisher himself said that a p-value of 0.05—a result sufficiently extreme that you'd only see it one time in every twenty—would be a good cutoff, although that was very much just a rule of thumb: "It is *convenient* to draw the line at about the level at which we can say: 'Either there is something in the treatment, or a coincidence has occurred such as does not occur more than once in twenty trials,'" he wrote: that is, $p = 0.05$. But that's completely arbitrary—you can and should select whatever level is most appropriate for the task at hand: "If one in twenty does not seem high enough odds, we may, if we prefer it, draw the line at one in fifty (the 2 percent point) or one in a hundred (the 1 percent point). Personally, the writer prefers to set a low standard of significance at the 5 percent point, and *ignore entirely* all results which fail to reach this level."[96]

Whatever level you decide upon is called your *alpha*. If your p-value is lower than your alpha, then you can "reject the null hypothesis" and treat the effect as though it's real. We call that *reaching statistical significance*. If your p-value is higher than your alpha, then we cannot reject the null, and we treat the effect as though it is not real.

TAILS I WIN

I've elided quite a lot of info in that IQ-and-feet example. What I did was called a "one-sample t-test," which compares the mean of a sample to a known population mean (in this case, IQ). As well as the sample size, you also need to know the standard deviation—in the case of IQ, fifteen. And you also need to decide whether it's a one-tailed or a two-tailed test.

Here's what that means. Imagine you're flipping a coin and you want to know whether it's fair or not. You flip the coin fifty times, and it comes up heads thirty-two times. What is the probability of something like that happening? Well, you work it out with your Pascal's triangle (or stick it in an online calculator, which is much easier), and it tells you: the probability of getting at least thirty-two heads from fifty coin flips is 0.03, or 3 percent. That's below Fisher's magical number of 0.05, so you can declare that it's statistically significant and get your "Coin Flipping: A Statistical Investigation" paper published in *Nature*.

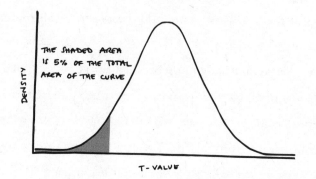

But hang on. Was there a particular reason why you thought the coin was biased toward heads? No? You'd have been equally surprised if it had come up with thirty-two tails, wouldn't you? So really you ought to be looking at the chance of a surprising result at either end of the spectrum. The probability of seeing thirty-two or more tails is also 0s03, so you can add them together—0.03 + 0.03 = 0.06.

The point is that unless you have some reason to only be looking at one end of the distribution, you'll be equally surprised by seeing extreme results at either end. So you need to look at both "tails." Which means that in order to be declared statistically significant, your result needs to be twice as extreme as it would need to be if you're only looking at one tail.

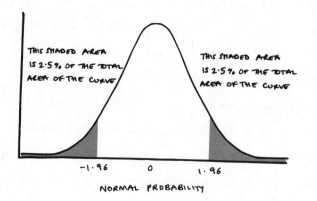

Of course, there's a lot more to it than this. But I think it's fair to say that this is the fundamental bit. You have a null hypothesis: there is no effect. You have an alternative hypothesis: there is some effect. If your data comes back, and it's sufficiently extreme that it would only appear one time in twenty or fewer given the null hypothesis, then (under Fisher's explicitly arbitrary ruling) you can reject the null hypothesis and act as though the alternative hypothesis is true.

Over the long run, in ideal circumstances, this should work. If you keep testing the IQs of lots of groups of people with big feet, and the null hypothesis is true that foot size and IQ are uncorrelated, then you'll only see weird outlier results one time in every twenty. And if you do see $p < 0.05$ results more often than that, then that's evidence that there is some correlation.

The trouble is, and we'll come back to this later, the circumstances are not ideal, and it is very easy to con yourself (or others) that a p-value of 0.05 or less means there is only a one-in-twenty chance that your hypothesis is false. In fact, it can be far, far more likely than that.

BAYES AT BAY

While the Bayesian model became unpopular, it never quite went away. For some things—as Fisher himself would acknowledge—it was the only way to do statistics. If you do know the background rate of some medical condition in the population, then it will give you the correct probability that someone has that disease if their test comes back positive; if you just go on the accuracy of the test itself, then you'll get it wrong an awful lot.

And despite Fisher's (and Neyman's and Pearson's) furious rejection of Bayesianism, people kept rediscovering or reinventing it, because it kept working.[97]

Harold Jeffreys, a Cambridge geologist, was the key figure in early scientific Bayesianism—he wrote that Bayes' theorem "is to the theory of probability what Pythagoras' theorem is to geometry."[98] While Fisher worked with pea plants and mice, experiments that gave precise answers and could be repeated as many times as required, Jeffreys looked at the propagation of seismic waves through the Earth. It was Jeffreys who first showed, in 1926, that the core of the Earth is liquid[99]—and, in fact, that the outer mantle is mainly silicon-based stone, while the inner core is mainly iron and nickel.

The data he used was much messier than that which Fisher was able to get his hands on. He wanted to use the time at which waves were detected at various seismological stations to pinpoint the epicenter of an earthquake, and also the nature of the material through which the waves passed. But earthquakes were relatively rare, and the data was noisy, so there was a lot of uncertainty. "Necessarily," wrote David Howie, a historian of statistics, "these inferences were tentative. They were advanced not with certainty but with degrees of confidence that were updated or modified to account for new information."[100] That is: they were done in a Bayesian fashion.

Each time Jeffreys got new information, he updated his prior confidence in his hypotheses. Jeffreys himself wrote: "Every scientific advance involves a transition from complete ignorance, through a stage of partial knowledge based on evidence becoming gradually more conclusive, to the stage of practical certainty." The uncertain parts of science, he said, are "the most interesting part."[101] He acknowledged uncertainty in everything, even—notably—for the correctness or otherwise of a scientific law. Dennis Lindley, another great Bayesian, wrote in a tribute after Jeffreys's death that "Jeffreys considered probability to be the appropriate description for all uncertainty, whereas statisticians usually restrict its use to the uncertainty associated with data."[102] That is, if you're not sure whether Geneva is the capital of Switzerland, or whether the universe is 13.8 billion years old, or whether your husband is cheating on you—all of which are things that are either true or false, whether or not anyone knows the true answer—Jeffreys would be happy using probability to express *your confidence* in the proposition.

(Howie also notes that Jeffreys was a "keen student of detective stories." In "fair-play" detective fiction, the reader should be given all the information that the fictional detective uses to solve the mystery. Devotees of this genre treat the novels more like a crossword puzzle than a story. Jeffreys would take notes on each character as he read, noting their alibis and motivations: "Another instance of drawing inferences from incomplete and unreliable data!" says Howie.)

Jeffreys sounds like a rather avuncular figure—"so quiet and gentle a man that it is hard to imagine his getting cross with anyone,"[103] says Lindley—who went from his childhood school in County Durham to St. John's College, Cambridge, and stayed there for seventy-five years, until his death. He contrasted oddly with his fire-breathing contemporary Fisher, but the two were friends of a sort, despite their deep philosophical disagreements. Jeffreys thought that the whole basis of frequentist statistics—the p-value and statistical significance, the "How likely is this data given the null hypothesis?" approach—was topsy-turvy.

He and Fisher engaged in a two-year debate in the pages of the *Proceedings of the Royal Society of London*, each arguing his case. It was inconclusive in substance, but Jeffreys lost in practice: frequentism remained the standard.

At around the same time, other scientists were trying to reconcile the problem of subjective priors. Three scientists came up with the same idea at around the same time as the Jeffreys-Fisher debate: Émile Borel, Frank Ramsey, and Bruno de Finetti. All three agreed that, yes, priors were subjective. But that didn't mean they were made up. Each scientist independently suggested that a way of quantifying priors was to place bets.

We'll go with Ramsey's version. Frank Ramsey was an English genius who, by the time he died at the age of just twenty-six, had already made major contributions to the fields of logic, mathematics, philosophy, and economics. His insight in probability theory was that probabilities are beliefs; our beliefs, if we act on them, are themselves a kind of bet. As Ramsey put it: "All our lives, we are in a sense betting. Whenever we go to the station, we are betting that a train will really run, and if we had not a sufficient degree of belief in this, we should decline the bet and stay at home."[104] This approach meant the bet could be quantified: the "probability of ⅓ is clearly related to the kind of belief which would lead to a bet of 2 to 1."[105]

This was the start of the idea of Bayesianism as decision theory, which we'll come back to. Ramsey's insight laid the groundwork for later work into economic rationality, and decision-making under uncertainty. "[He] set out a framework that tells us what is rational, given an agent's beliefs and desires," wrote Ramsey's biographer Cheryl Misak.

Say you are at a crossroads, unsure of which path will get you to the parking lot most quickly. Say that, if you choose the shorter route, you will get 30 units of happiness or well-being, and if you choose the longer route, you will get 18 units. You have a degree of belief of ⅔ that the right-hand road is the shorter route, and ⅓ that the left-hand road is shorter. Ramsey's model lets you calculate your expected subjective utility for each option, and tells you that, given your beliefs and desires, it is rational for you to take the right-hand road.[106]

So your priors are subjective, true—but they can still be good or bad, and they can be tested. More than that, they can be incoherent. If your probabilities don't add up, then you can be taken for a ride by anyone offering you bets. Imagine a bookie that offered odds on three horses: one at evens (you'd double your stake if you won), one at three to one against (quadrupling your stake), and one at four to one against. If you took the bookie up on all three bets, then you could bet £100 on the first, £50 on the second, and £40 on the third. You'd have paid £190 in bets, but you'd be guaranteed £200 in returns. If probabilities are beliefs, and beliefs are implied bets, then your beliefs—though subjective—can be inconsistent in the same way.

Away from science and professional statistics, Bayes' theorem continued to burble away. During the Second World War—when, among scientists, Fisherian frequentism was dominant—people with more urgent, practical concerns over how to make good inferences from limited data were either using Bayes' theorem or developing it themselves.[107]

Alan Turing, the great Cambridge mathematician, was brought into the British military during the war to help crack German codes and stop transatlantic shipping getting sunk by U-boats. The U-boats communicated with their bases via radio, but the messages they sent were encrypted via a mechanical device known as the Enigma machine, which created new ciphers regularly. Turing built an early computer to try to decode it—to work out which letter was standing for which.

The trouble was that there were an extraordinary number of different possibilities. Treating them each as equally plausible would have meant that, if the machine checked one hundred possible combinations a second, it would have taken many trillions of times the life of the universe to check them all. It meant that Turing and his team had to use priors—to assume that some combinations of letters were more likely than others. So the three-letter sequence E-I-N, *ein*, German for "a" or "one," would be more likely—he reasoned—than the sequence J-X-Q. Words like "WIND" or "CONVOY" would be more likely than "ITCH" or "BALLET."

Turing formalized this intuition into a Bayesian framework and built the beginnings of modern information theory around it—he created a term called the "ban," which referred to a unit of information comparable to the modern bit or byte.

At the same time, insurance underwriters were using Bayesian methods to set premiums for workplace liabilities. Quality-control assessors in military manufacturing were using Bayesian ideas to minimize the number of shells that had to be tested to establish confidence. Artillery commanders used Bayes' rule to give the best firing solutions. But in science, frequentism still ruled, until the 1970s.

MINE EYES HAVE SEEN THE GLORY

"I only ever went to one," says Andy Grieve. "After that one my wife wouldn't let me go any more, because I missed the plane coming back and wasn't there for a wedding the following weekend. They were unbelievable."

Grieve is a statistician, a former president of the Royal Statistical Society, in fact, now semiretired after nearly five decades in the pharmaceutical industry. He's also a Bayesian. He's talking about the legendary Valencia conferences, where Bayesianism reached its modern form.

In the 1970s, the great statistician Dennis Lindley, then the head of the Department of Statistics at University College London, had—according to the Spanish mathematician José-Miguel Bernardo, who had just completed his PhD there—turned his fiefdom into "*the* European Bayesian department"[108] of the era.

That might not have been as huge an accolade as it sounds: away from UCL, Bayesianism was very much a sideshow. "At University College the world looked Bayesian; thus, it came as a kind of a shock to discover that in most statistical conferences you had to fight for your right to work within Bayesian statistics to a mainly unsympathetic audience, with no real time left to go into the details of your work," Bernardo went on. Grieve remembers something similar: "When we were first giving public lectures on Bayes," he says, "it wasn't unusual to be given the before- or after-lunch slot. The comedy slot. 'Andy's going to be talking about Bayes.'"

In 1976, Lindley, Bernardo, and the by now somewhat elderly Bruno de Finetti attended what they believed to be the world's first international Bayesian conference, in Fontainebleau, France. Over a pleasant lunch, having spent several days not having to start every conversation with an argument about Bayesianism versus frequentism or being treated like the "... and finally" item at the end of the BBC's *The Six O'Clock News*, the three of them decided that they ought to do it again. A similar experience was had in Florence a year later, and then Bernardo took a position at Yale and spent some time traveling the United States giving seminars, where he met a great selection of brilliant thinkers—George Box (son-in-law of Ronald Fisher) and I. J. "Jack" Good (Bletchley Park veteran and early AI theorist) among them. After one of these seminars, Bernardo got chatting with the statistician Morris DeGroot, and "during a very long evening, with plenty of scotch, we talked about many aspects of life and somehow, by dawn, we came to talk about statistics, and we agreed to make an effort to try to organize an international Bayesian meeting at the first available occasion."

Bernardo then took a professorship in biostatistics at the University of Valencia, just as Spain was emerging from decades of fascist dictatorship and starting to open up. He suggested to the Spanish education minister that they make Valencia the venue for "a first Bayesian world meeting." The first one was in 1979.

I say this cautiously, because people probably won't believe it about an academic statistics conference, but it sounds pretty wild. "They were a lot of fun," Grieve says, with a tone of wistful reminiscence. "A lot of hard work, but a lot of fun. We'd work in the mornings, then have a siesta—2 p.m. to 6 p.m. we'd do nothing in terms of statistics—and then 6 p.m. to 10 p.m. working again, then we'd have dinner. The parties where we used to sing didn't start until 10 p.m."

Yes, the parties where they used to sing. After dinner at the first conference, Bernardo recalls, "George Box sang 'There's No Theorem Like Bayes' Theorem,' a parody of 'There's No Business Like Show Business' by Irving Berlin."[109] This established a tradition called the Bayesian cabaret. A few minutes on the internet will find you more examples than you could possibly need. There exists, stored on the servers of the University of Minnesota, for reasons I don't fully understand, a "Bayesian songbook," containing such wonders as:

Thomas Bayes's Army [The Battle Hymn of Las Fuentes]
Words: P. R. Freeman and A. O'Hagan
Music: Traditional ["The Battle Hymn of the Republic"]

Mine eyes have seen the glory of the Reverend Thomas Bayes,
He is stamping out frequentists and their incoherent ways,
He has raised his mighty army at the Hotel Las Fuentes,
His troops are marching on.
Glory, Glory, Probability
Glory, Glory, Subjectivity
Glory, Glory, on to infinity
His troops are marching on!

It continues in this vein at some length.[110]

Over the years there was also a song called "José Bernardo," which was sung to the tune of the Macarena; Andy Grieve sang a repurposed medieval students' drinking song, "*Gaudeamus Igitur*," along with another future president of the Royal Statistical Society, Professor Sir David Spiegelhalter; there was a "Bayesians in the Night" to the tune of "Strangers in the Night"; a "Like a Bayesian" ("Like a Virgin"). And so on.

I mentioned this on Twitter and Sir David got in touch to say that, alas, "Our performance of 'The Full Monty Carlo' was before the smartphone era, so no recordings exist."[111] ("Who would want to see a video of six male professors of Bayesian statistics taking their clothes off in front of a screaming crowd in a Spanish nightclub?"[112] he went on to ask, in my view entirely misjudging the nature of the modern internet.) There absolutely *are* videos, from later conferences, of "Bayesian Believer," a Monkees reimagining ("Then I saw Tom Bayes, now I'm a believer"), and "What a Bayesian World," à la Louis Armstrong.

Another attendee of the second Valencia conference told me that he, José Bernardo, and a cohort of the world's most eminent Bayesian statisticians went for a swim off a boat up the coast, the wind nearly blew them onto some rocks, and it all got a bit hairy and life preservers were thrown. "If we'd all have drowned," he said, "that would have knocked the development of Bayesian statistics on the head rather."

"I have a sweatshirt from the third Valencia conference," Grieve says cheerfully, "saying 'Bayesians have more fun.'"

In the decades since Fisher and Jeffreys fought it out, and since Bayesian methods were all but removed from scientific work, even as they were quietly used almost by default in other areas, Bayesianism has made something of a comeback. Partly that was Jeffreys's book being passed down almost like samizdat. ("Ah, you want to calculate the probability that a hypothesis is true, not just the probability that you'd see this data if it is true? Here, try this.") Partly it was that in a lot of areas—software engineering, most notably—a form of Bayesianism just kind of fell out of the way people did the numbers, as Turing had found some decades earlier. "There's an interesting dynamic," Aubrey Clayton says. "People who come from the new schools of data science—machine learning, Silicon Valley tech folks—they see this as more of a settled debate, for the same reasons Turing did. You do these problems and of course Bayesian methods are what you'd use. Outside of academic science, you might get the idea that Bayesianism is what everyone uses." [113]

Again, Grieve says something similar: "About the turn of the millennium, I was working in Connecticut for Pfizer, and I spent the weekend visiting an old friend, an electronics engineer who used to work for Hewlett-Packard. He'd set up his own business and was developing methods for searching through data quickly on large computer discs. The algorithms he'd developed were essentially Bayesian. He was totally unaware of what he was doing. There are lots of areas where Bayesian algorithms are being developed and people don't realize."

As Bayesian methods became more widespread, tensions started to rise. Bayesians were the underdog and also the up-and-comers, frequentists the establishment. (Bayes himself, a Nonconformist outside the established Church, might have appreciated the irony.) It got remarkably heated, at least in public debates.

"There was a famous statistician, Maurice Kendall," says Grieve, "whose textbook[114] on statistics was around when I was a student. He wrote a paper in 1968 in which he said that if Bayesians would only do as Bayes did and publish posthumously, we'd all be saved a lot of trouble."[115] Meanwhile Dennis Lindley wasn't exactly pouring oil on troubled water, telling a conference in 1975 that "the only good statistics is Bayesian statistics. [Bayes] is not just another technique to be added to our repertoire alongside, for example, multivariate analysis: it is the only method that can produce sound inferences and decisions."[116]

There was a certain bright-eyed evangelical quality to Bayesians, as there often is when a small group outside the mainstream believes that it has the truth (see also: the environmentalist movement, cyclists, literal Evangelists). That doesn't mean that they're wrong—you can't tell whether something's true or not by psychoanalyzing the people who believe it. But it does mean that Bayesians could be quite annoying, and as a result, they annoyed the frequentist establishment.

It also meant that they often got called a "cult," a situation that may not have been helped by the singalongs about "Mine eyes have seen the glory." The statistician Larry Wasserman published a blog post titled "Is Bayesian Inference a Religion?" in 2013[117] (his answer was that for a subset of Bayesians it was), and an anonymous former Bayesian statistician responded to it saying that "I used to be one of those believers in the Bayesian Truth," but he had "lost my faith."[118]

We shouldn't overstate the enmity between Bayesians and frequentists—especially nowadays, but even when the Valencia conferences were going on and Bayesianism was growing in confidence. Grieve remembers that after roaring debates at the Royal Statistical Society, "they'd slag each other off in public, in those arguments and disputes. But afterwards they'd be going to dinner in a taxi together. While it does sound tough, a lot of them were great friends." He remembers reading a paper by Spiegelhalter, a fellow Bayesian, in the 1990s, and thinking, "They had joined a recent tradition of avoiding controversy and pursuing the practical benefits, even though the former was more entertaining. I do think there was a bit of entertainment in it."

There's a term in professional wrestling, *kayfabe*, which means maintaining the illusion of reality—so the Undertaker still goes on about how much he hates Hulk Hogan even once the bout is over and he's just doing an interview or even walking to a store. There seems to have been an element of kayfabe in the Bayesian-frequentist wars. Grieve points out that George Box, the first of the Valencia cabaret artists, wrote a paper in 1985 essentially admitting as much. In "An Apology for Ecumenism in Statistics," he says, "I believe that scientific method employs and requires not one, but two kinds of inference," Bayesian and frequentist.[119]

This does seem to be the case. Jens Koed Madsen, a cognitive psychologist at the London School of Economics, told me that he uses both frequentist and Bayesian statistics, depending on the question he's trying to answer. Sophie Carr, a Bayesian statistician and founder of a consulting firm literally called Bays, says that a lot of the work she does isn't Bayesian at all. And Daniël Lakens, a statistically minded psychologist who has a reputation as an arch-frequentist and hammer of the Bayesians, cheerfully admits in his online course that you need

Bayesianism if you're going to make statements about how plausible a hypothesis is, and told me when I spoke to him that Bayes really is necessary for decision theory.

Perhaps now we ought to take a look at the state of statistics in science today.

CHAPTER TWO

Bayes in Science

THE REPLICATION CRISIS IN SCIENCE, AND SOME WAYS TO FIX IT

In 2011, a series of unwelcome things happened, and science was shaken to the core.

Not everyone noticed. The "replication crisis," as it was known, probably didn't affect your daily life (it didn't affect mine for some years, and I was writing about science for a living). Most scientists—even most psychologists, whose discipline was the worst affected—were able to go on for quite a long time as if nothing had happened. But 2011 was a very important year for science, and for scientists who had even a vague understanding of statistics—and who cared about finding out true things rather than just getting citations and tenure—it has a certain Year Zero quality to it.

First, a senior scientist was found to have been fraudulently making up his data. Diederik Stapel, a rising star in social psychology and a professor at Tilburg University in the Netherlands, had made a splash with a series of headline-grabbing papers: one suggesting that eating meat made people more antisocial;[1] another finding that people are more likely to be racist if the environment they're in is filled with litter.[2] But it turned out that for those two studies and several others, he had never performed the experiments or gathered the data. He had just made it up.[3] That's not great, but sometimes fraud happens. It was detected (eventually). He was fired and dozens of his papers were retracted from the scientific record.

What was more worrying, for thoughtful scientists, was how science could go wrong even when scientists weren't committing fraud.

In March of the same year, the social psychologist Daryl Bem, working at Cornell University in New York State, published a study called "Feeling the Future."[4] It was a classic social psychology study, in many ways, known as a "priming" study. In priming, experimenters take a bunch of subjects—usually university students being paid a few bucks or given course credit—"prime" them with some concept, and see how it affects their behavior. The priming might be, for instance, that you give them some jumbled-up words to unscramble, and then make them do a task, and see if the words you've primed them with affect how they do that task.

The study of priming started out with relatively unremarkable things, like if you give someone the word "doctor" to unscramble, then afterward they'll be faster at recognizing the word "nurse" than they are at recognizing unrelated words. But then they started investigating more surprising things: the idea that priming in these subtle ways had major, dramatic effects on our behavior.

A very famous example, for instance, was John Bargh's 1996 study,[5] which found that priming people with words like "wrinkle," "bingo," and "Florida"—words associated with age, especially in the US—made them walk more slowly as they left the laboratory, as though they had aged. Another, from 2006, found that priming people with concepts related to money made them less willing to seek or offer help—to be more selfish, basically.[6] Another found that exposing people to the smell of fish made them more suspicious, because—seriously—it "smells fishy."[7]

These studies were the bread and butter of social psychology in the 1990s and 2000s. They'd been around since the late 1970s, but their heyday came a few decades later. And they seemed to demonstrate an extraordinary susceptibility in the human mind: we could be manipulated into all sorts of strange behavior by subtle, unconscious cues. Those people walking slowly out of John Bargh's laboratory had no idea that the word "Florida" had made them think of shuffleboard-playing retirees, but somehow it had, and more than that, it had made them behave like one. It seemed to show that our conscious minds were just blown around, leaves on the wind of almost undetectable cues from our environment. You may not have heard of it, but you've probably heard of its intellectual offspring, "nudging" and "subliminal advertising." When Tyler Durden in *Fight Club* splices frames of pornography into children's movies, too fast for people to see with their conscious minds, it's based on ideas taken from priming research.

In his book *Thinking, Fast and Slow*—also published in 2011—the Nobel Prize–winning psychologist Daniel Kahneman wrote of priming: "Disbelief is not an option. The results are not made up, nor are they statistical flukes. You have no choice but to accept that the major conclusions of these studies are true."[8]

But then along came Bem.

Bem's study consisted of several experiments, but we'll focus on just one of them: an entirely unremarkable example of priming in all ways except one. The experiment, like the rest of them, gave a prime and saw how it affected behavior. In this case, the subjects were primed with a positive or negative word ("beautiful," say, or "ugly") and then were shown images and asked to press a button, as quickly as possible, to indicate whether the image shown was pleasant or unpleasant. Traditional priming literature would predict that people given a positive prime would be quicker to say that pleasant images were pleasant, slower to say unpleasant images were unpleasant, and vice versa.

The big twist, though, was that in half of the trials, the subjects were given the prime *after they had been shown the image*. And—this is the important bit—the priming *worked*. People were quicker to indicate that pleasant images were pleasant when they were given a positive word, even though the word didn't appear until after they'd made their choice.

The finding was statistically significant—a p-value of 0.01; enough, by modern convention, to reject the null hypothesis. And Bem suggested that this was evidence for "psi"—for psychic powers, clairvoyance. The other eight experiments in the study, using subtly different methods, but all of them essentially social-psychology staples with the time order reversed, similarly achieved significance.

Most of us would probably agree that psychic powers don't exist. But here was an apparently well-carried-out study that appeared to find—nine times!—that they do. It used the same methodological and statistical tools as other psychology studies; it used the same cutoff for significance, of p = 0.05. So either clairvoyance is real, or there was something going wrong with statistical practice. (Daryl Bem, for the record, still believes that the answer is that clairvoyance is real and that his studies were correctly detecting it.)

The third great blow of 2011 came in the form of a paper called "False Positive Psychology" by the psychologists Joseph Simmons, Leif Nelson, and Uri Simonsohn.[9] It did much as Bem's paper did—used bog-standard statistical techniques to prove an impossible result. But unlike Bem, Simmons, Nelson, and Simonsohn did it on purpose, to show that those bog-standard statistical techniques, used throughout science, were badly flawed.

Again, there were various experiments within the study, but we'll focus on the most famous one. In the study, they asked twenty under-graduates to listen to a song—either "When I'm Sixty-Four" by the Beatles, or "Kalimba" by Mr. Scruff. Then they compared the ages of the two groups. It turned out that people who had listened to "When I'm Sixty-Four" had become nearly eighteen months younger. Again, it was statistically significant: p = 0.04.

Once more I think most people would agree that it is unlikely that listening to the Beatles actively makes people younger—not simply makes them feel younger, but makes their birthdays become more re-cent. The result cannot be real. And yet, once more, the "False Positive Psychology" paper proved it to be real, to the standards of modern social science, and only used the same statistical methods that other scientists were using every day.

Some scientists had been warning that something like this was coming. In 2005, Stanford's John Ioannidis had written a paper titled simply "Why Most Published Research Findings Are False."[10] It said that the statistical practices in much of science left it open to this sort of problem. The problems in science were many and varied, but a major one was that scientists weren't asking how likely their hypothesis was to be true, given the data they had collected—they were asking (as Bernoulli had, and Fisher) how likely it was that they would see the data they had collected, if the hypothesis was false.

Funnily enough, the apostle of modern Bayesianism, Dennis Lindley, had foreseen something of this problem way back in 1991, when he wrote a tribute to Harold Jeffreys in the journal *Chance*. "Many experimentalists, when asked what 5% significance means, often say that the probability of the null hypothesis is 0.05," he wrote. But, of course, that's not what it means: it's just how likely you would be to see data at least that extreme, if the null hypothesis *were* true. In fact, Lindley pointed out, if you used Jeffreys's Bayesian methods and agreed to only publish if there were a 5 percent chance of the null hypothesis, then you would reject a lot of published papers: "It is more likely to have significance at 5%," he wrote, "than to have a probability as low as 5% for the null hypothesis. Thus, on this scale, a significance test is more likely to suggest a difference than is Jeffreys's method."[11]

And then he put his finger on the crux of the problem. "This may partly account for the popularity of tests with scientists, since they often want to demonstrate differences," he wrote. "It would be interesting to know how many significant results correspond to real differences."

The answer, the scientific world would learn to its horror twenty years later, was "nowhere near as many as you'd hope."

How can that be, though? If a study finds a p-value of 0.05 or less, that means that you'd only see those results (or more extreme ones) by chance one time in twenty at most, doesn't it? So surely, if every study is using that yardstick, you wouldn't expect to see many false positives.

That is the idea, sure. But it's not as straightforward as that. The easiest way to get a $p < 0.05$ result—that is, something that you'd only see by coincidence one time in twenty—is to do twenty experiments, and then publish the one that comes up. That's exactly what the "False Positive Psychology" people did: they measured lots and lots of things, when they were looking at their undergraduates. Their parents' birthdays, how old they *felt*, their political orientation, whether they referred to the past as "the good old days," a whole bunch of things. They also gave them another song to listen to: "Hot Potato" by the Wiggles.

Then they cut their data up in different ways. Are people who listened to "Hot Potato" more right-wing than people who listened to "Kalimba"? Did "Kalimba" make people more nostalgic than "When I'm Sixty-Four"?

If you chop these things up in different ways, with a small sample size like twenty, you can easily get false positives. They also did other things, like stopping collecting data if their p-value dropped for a moment below 0.05. Simmons, Nelson, and Simonsohn estimated that by running a few simple tricks like this, you could make it more than 60 percent likely you'd find an apparently significant result.

This is known as "hypothesizing after results are known"—HARK-ing—or "p-hacking," and it happens all the time, not just in wry papers intended to demonstrate that it's possible. One example: there's a thing called the competitive reaction time task (CRTT), which is used to measure aggression, especially in research into the psychological effects of video games. A player plays either a violent or a nonviolent video game. Then they take part in a competitive game against an opponent: the first to react to some stimulus wins. And the winner gets to blast their opponent with some noise, potentially at a painful level.

The twist is that the opponent in the game isn't real—it's just a computer program. But Malte Elson, a psychologist, noticed that in the 130 papers that had been published using the CRTT in video game aggression research by 2019, the data had been analyzed 157 different ways.[12] Sometimes they measured volume of the first blast, sometimes average volume over twenty blasts, sometimes duration of the first blast, sometimes volume times duration, and so on. It would be near-impossible not to find a significant result like that.

You might ask: Why? Why would people do this, if they're trying to find out whether something is true or not? The answer, at least in part, is that scientists, although they *do* want to find out whether things are true, also want to be promoted, and get tenure, and feed their families, and all those boring things. The basic driver of academic success is summed up in the phrase "publish or perish." If you're not getting your research published regularly in journals—preferably "high-impact" journals like *Nature* and *Science*—then you're not going to get that professorial position at a redbrick university.

This wouldn't matter, so long as journals published every study that was submitted to them, regardless of whether or not the researchers found whatever they were looking for. But, of course, they don't.

Science journals—not *all* science journals, but most, including most of the big names—publish results that are interesting and novel. That might not sound too terrible, but it means that a study that finds something interesting—"psychic powers are real," for instance—is more likely to be published than a study that finds something more boring, like "we looked for evidence of psychic powers and didn't find any."

And, of course, a lot of journals—again, not all, but a lot, especially in the social sciences—use $p < 0.05$ as their threshold for "found something interesting." If your experiment returns results that are $p = 0.045$, it may well get published. If it returns results that are $p = 0.055$, it may well not.

This is a problem for science in its own right. Imagine one hundred labs carry out studies into whether or not psychic powers are real, and ninety-five of them find nothing, but five of them find statistically significant results ($p < 0.05$! You'd only see results like that five times out of every one hundred if it wasn't real!). But because journals want to publish interesting, novel things, they might very well publish all five of the "psychic powers are real" papers and only one of the "psychic powers aren't real" papers, meaning that if someone went to the scientific literature, they'd find that 85 percent of studies looking into psychic powers find them. If you speak to many scientists, you'll hear a lot of stories of them getting rejections because their results weren't "novel" enough, which, of course, means that the scientific literature systematically fills up with "novel," exciting studies that do find things, while the boring, not novel, but often more *actually true* findings are rejected.

But it also has knock-on effects, in that it incentivizes scientists to find ways—even if only subconsciously—of getting that $p < 0.05$ positive result if they possibly can: often by doing the exact things that the "False Positive Psychology" people did so elegantly.

Perhaps the most famous example of this was the food scientist Brian Wansink, a star at Cornell University who received millions of dollars in US federal government funding under the Obama administration. He published lots and lots of studies about our eating behavior—notably, one about how men eat more in the company of women[13] (presumably to impress them); another about how giving vegetables more "attractive" names (calling carrots "X-ray vision carrots," for instance) makes elementary school children eat twice as much of them.[14]

Then, in 2016, he made the mistake of publishing a blog post titled "The Grad Student Who Never Said 'No.'"[15]

The grad student in question was a Turkish PhD candidate. When she arrived at Cornell, Wansink "gave her a dataset of a self-funded, failed study which had null results"—a study that looked at eating behavior in an all-you-can-eat Italian buffet over a month. In his words, he told her: "This cost us a lot of time and our own money to collect. There's got to be something here we can salvage because it's a cool (rich & unique) data set." So the PhD student went off and cut up the dataset in lots of different ways. And, inevitably enough, she found lots of $p < 0.05$ correlations—enough for her and Wansink to publish five papers from it (the "men overeat to impress women" paper among them).

This raised some scientists' and science journalists' eyebrows, and they started raking through Wansink's other research. Also, Stephanie Lee, a science journalist at *BuzzFeed*, got hold of his emails, in which—it transpired—he had told his PhD student to cut the data up into "males, females, lunch goers, dinner goers, people sitting alone, people eating with groups of 2, people eating in groups of 2+, people who order alcohol, people who order soft drinks, people who sit close to buffet, people who sit far away, and so on," in order to "mine it for significance . . . squeeze some blood out of this rock" and get it to "go virally big time."[16]

As a result, eighteen of Wansink's papers have been retracted; seven have received "expressions of concern," which journals append to studies they don't think can be fully trusted, but aren't ready to retract altogether; and fifteen have been corrected.[17] Wansink, meanwhile, resigned from Cornell in 2019, after the university found him to have committed scientific misconduct, and barred him from teaching and research.[18]

This is a particularly egregious example, but in a way Wansink was unlucky that he was publicly destroyed for something that was almost standard practice. P-hacking goes on all the time, in much less dramatic ways—and a lot of scientists have absolutely no idea that they're doing anything wrong. The aforementioned Daryl Bem, in a 1987 book chapter written as a guide to help students get their research published, wrote that "there are two articles you can write: the article you planned to write when you designed your study; the article that makes the most sense now that you have seen the results. The correct answer is the second one."

He called for researchers to "analyze the sexes separately, make up new composite indexes . . . reorganize the data to bring them into bolder relief. . . . The data may be strong enough to justify recentering your article around the new findings and subordinating or even ignoring your original hypotheses. . . . Think of your dataset as a jewel. Your task is to cut and polish it, to select the facets to highlight, and to craft the best setting for it." [19] It's not intended as a call for p-hacking, but "recentering your data around the new findings" is exactly what both the "False Positive" guys and Wansink were doing, and as they demonstrated, if you do that, you can very easily get statistically significant findings from utterly meaningless noise.

So far we've just looked at specific scientists, but it's worth getting a sense of how big a problem this was in science at large. In late 2011—alarmed by the various things that had come to light that year—Brian Nosek, a psychologist at the University of Virginia, launched something called the Reproducibility Project. He got 270 researchers to collaborate in attempts to replicate one hundred psychology studies—that is, redo the experiments, using the same methods but new data, and see if they found the same results.

Nosek and his collaborators published the resulting paper in 2015. [20] Of the one hundred studies they looked at, ninety-seven had originally found statistically significant results; Nosek et al. were able to do the same in just thirty-six. The effect sizes of the replications were, on average, half the size of the originals. More than half of those effect sizes fell outside the 95 percent confidence intervals of the original papers' findings. Ioannidis's (and Lindley's) warnings—that many, possibly most, scientific findings in the published literature were false—had been shown to be prophetic.

You might be wondering what this all has to do with Bayes' theorem. Well: the cause of the replication crisis has been greatly discussed. It is a story of bad incentives—publish or perish, the demand for novelty—and scientists have come up with many sensible proposals for how to fix them. Lowering the threshold for "significance" is one; requiring preregistration of hypotheses in order to prevent HARKing is another; having journals agree to publish papers on the strength of the methods, not the nature of the findings, in order to avoid the novelty filter, is a third.

But you could go deeper and say that the underlying cause of the replication crisis is even more basic: it's that science, like Jakob Bernoulli three hundred years ago, is doing sampling probabilities, not inferential probabilities.

A p-value is not, as we've discussed, a measure of how likely it is that your hypothesis is true, given your data. It's a measure of how likely it is that you would *see* that data, given a certain hypothesis. But—as Bayes noted, and as Laplace later fleshed out—that's not enough. If you want to measure how likely it is that your hypothesis is true, you simply cannot avoid priors. You need Bayes' theorem. The question, of course, is whether that *is* what you want.

A MOON MADE OF CHEESE, PSYCHIC POWERS, AND FASTER-THAN-LIGHT PARTICLES

Here's the fundamental point, as Bayesians would see it. Imagine you do some study to test some hypothesis—we won't say what it is yet—and you get a p-value of 0.02. How likely is it that your hypothesis is true? It's an annoying fact that a large number of people who *definitely should know better* would say that the probability is 98 percent. The chance of seeing those results by chance is one in fifty, so the probability that it *is* coincidence is 2 percent, right? Hopefully by this stage you'll know that that's not true. But most scientists, it seems, don't.

A 2007 study asked forty-four psychology undergraduates, thirty-nine psychology professors, and thirty psychology professors who *specifically act as instructors in statistical methods* to read six statements about statistical significance and mark them as true or false.[21] Every single undergraduate, 90 percent of the professors, and 80 percent of the methodology instructors—again, the *people whose job it was to instruct students on statistical methods*—marked at least one statement wrong. A third of the two latter groups, and two-thirds of the undergraduates, thought that the p-value indicated the probability that your results were due to chance, given the data—the probability, that is, that the null hypothesis is true. So if you have a p-value of 0.05, that means there is only a one-in-twenty chance that your hypothesis is false. That is, of course, not the case.

What's perhaps even more astonishing than that is that another study looked at thirty Introduction to Psychology *textbooks* and found that twenty-five of them gave a definition of "statistical significance," and that twenty-two of those twenty-five were wrong.[22] Again, the most common error was that they assumed a p-value gave the probability that the results were due to chance. This is (as we've been discussing for quite some time now) completely backward. What p-values tell you is how likely you are to see that data, given a hypothesis.

But the fact that not everyone understands what p-values are for doesn't mean that everyone who advocates for them is an idiot or that they don't understand them.

I spoke to Daniël Lakens, a psychologist at Eindhoven University of Technology in the Netherlands and, in the circles I move in, largely considered an arch-frequentist.* He cheerfully acknowledges that, unless you know the prior probability of your hypothesis, you can't know from the p-value how likely it is that your hypothesis is true.

What a p-value does, he says, is let you know how often, over the long run, you *would* get false positives if the null hypothesis is true. And, he says, if you get a p-value of less than 0.05, then that allows you to *act as if* you have rejected the null hypothesis—to do more research, or to publish your study. But it's only ever provisional.

The trouble Bayesians have with this framework is that it privileges really stupid ideas. Go back to that study, with the p = 0.02 finding. Let's say it was a study into whether hammers fall downward faster than helium balloons (in Earth's ground-level gravity and atmosphere). You drop one hammer and one helium balloon together six times, and you find that the hammer hits the ground first each time. Your p-value for that result, on a one-tailed test, is about 0.02. It's statistically significant! You're pretty unlikely to see that result by chance. Hooray. But it's not really very exciting.

But now let's say that, instead, the study was into the existence of psychic powers. You get some undergrad to choose between two identical pictures, but *after* they choose, you flash a pornographic image where one of them used to be (one of the experiments from Bem). You do this six times. All six times, the undergrad chooses the one that is followed by the porn pic. Again, your p-value is about 0.02.

---

* I strongly recommend Lakens's free Coursera course online, "Improving Your Statistical Inferences," if you want to know more about this stuff.

As far as the frequentists are concerned, that's all the information you have to go on. You have your data, you have your hypothesis. Neither is any better than the other.

According to the frequentist model, you're justified in treating both of those findings the same—as permission to *act as if* the null hypothesis is false and there is a real effect here. But most of us would probably agree that there genuinely is a real effect in the hammers-fall-faster-than-helium-balloons study. Your $p \approx 0.02$ result doesn't really change that very much. You believed it already. And most of us probably agree that there *isn't* a real effect in the "undergraduate students can psychically detect porn" study. If it's true, it's really surprising.

Given that journals want novel, exciting results, and given that frequentist models don't take into account prior probabilities, and given that if you do twenty experiments the odds are you'll get a statistically significant result in at least one of them even if there's no real effect, there's an obvious incentive to do the "Are undergrads psychic?" experiment. "If everything is being assessed with the same rubric," says Aubrey Clayton, "you might as well choose the most outlandish theory you can because it'll get the most buzz and notoriety. With frequentism, people have incentives to come up with the novel, surprising theories."

He argues that, instead, you should take prior probabilities into account. "If you have some hypothesis, like 'the moon is made of cheese,' you'd have very low priors, so new data doesn't move the needle much," he says. "It might give you a hint of being convinced, but it doesn't drown out your prior skepticism. That's what Bayesian statistics gives the scientist, a vehicle for skepticism, a way to say, 'I don't believe this theory.'

"It's a perverse incentive, for scientists to be looking at theories that we should have a priori skepticism about. We should raise the evidentiary bar."

For Lakens, though, this is silly. "The Daryl Bem stuff is the perfect example," he says. "Popper [Karl Popper, the great twentieth-century philosopher of science] talks a lot about dogma—he doesn't want dogma to enter the scientific process.

"So if an editor says, 'I don't believe this precognition stuff,' I'd say, 'I don't care. Shut up. Publish the things.'"

He gives another, perhaps more important, example. In 2011, an experiment at the European Council for Nuclear Research (CERN, best known for the Large Hadron Collider) spotted something extraordinary.[23] CERN had a particle accelerator in Geneva and a particle detector in Italy, and the former fired neutrons toward the latter, 730 kilometers (450 miles) away. Using atomic clocks, precise to some unfathomably tiny fraction of a second, the researchers recorded the time the neutrons left the accelerator and arrived at the detector.

They noticed that the neutrons arrived in Italy sixty-billionths of a second sooner than they thought possible. That was an *extremely* statistically significant result—it had a p-value of about 0.000000002, meaning that if it was purely by chance, you'd only see results that extreme one time in every 500 million.

It was also extremely unlikely to be true. By arriving those crucial sixty nanoseconds too soon, the neutrinos had apparently broken the speed of light. Nothing can break the speed of light—it is perhaps the most fundamental axiom of the relativity theory. As things get faster, their mass increases, and that mass approaches infinity as they near light speed. In order to get any particle with any mass at all to the speed of light would require infinite energy, which is impossible. If this CERN finding was real, it would involve a huge rewriting of modern physics.

So even given a p = 0.000000002 result, most physicists would have been extremely confident that the finding was not, in fact, real. "The result is impossible," says Lakens. "But should we have hidden it, because it was an embarrassment? No. You don't hide it, that's not how it works. And sometimes in history [these weird findings] have been the breakthrough thing that we didn't want to miss. I don't like dogma in science."

As it turned out, of course, the CERN result wasn't real. Further investigation found that a fiber-optic cable in the clock system had not been screwed in properly—this meant that a laser signal within the clock was not picked up quite so quickly, speeding the apparent arrival of the neutrinos by about 75 nanoseconds—enough to make it seem as if they had got there before a beam of light could.[24]

Perhaps we should point out that, in a way, the frequentist-Bayesian argument doesn't matter here, or at least it's more complicated than I've made it seem. Whatever your Bayesian prior is, unless it's something outrageous, then a six-sigma (standard deviation), p = 0.000000002 result would overwhelm it easily. I don't think a reasonable Bayesian would be so confident as that: they'd need to think there's only about a one-in-several-hundred-million chance that light speed could be broken. *If you believed that the only explanation for the neutrinos' apparent early arrival was that they had genuinely traveled that fast*, then the result should have convinced you.

But no one did think that. So it wouldn't have mattered how strong that effect was, because no physicist would have believed that it meant that the theory of relativity had been overturned. Instead, they'd have assumed that there was some other explanation—a measurement error, an equipment fault, perhaps fraud. And of course there was (not fraud, I hasten to say). The result wasn't chance—there was a real effect, but it was caused by the faulty fiber-optic cable, not by superluminal particles.

Later on, we'll talk about what happens when you get strong statistical evidence for some highly implausible theory—often, the good Bayesian thing to do is to assume that the evidence is misleading in some way. (Which gets controversial.)

Anyway. Lakens's point is that you don't get to pick and choose. If a scientific experiment is seemingly well performed, and it returns some startling result, then, he says, it's not right that you simply don't publish it on the grounds of having a low prior. He cites Popper again: "Popper hated Bayes. He didn't want Bayes as part of his philosophy of science. I'm just joining him in saying it." Popper's philosophy of science said, to oversimplify, that you never prove a scientific hypothesis—you only disprove it or fail to do so. The Bayesian idea that you can build up evidence for or against is very much opposed to it.

Lakens, in fact, disagrees with the entire premise of the Bayesian revolution—or the foundational tenet of the Bayesian faith, if you prefer. The line I have been repeating is that frequentist statistics tells us, "How likely are we to see this data, given this hypothesis?," but that *what we really want to know* is "How likely is it that my hypothesis is correct, given this data?" Lakens rejects this entirely. "I call this the statistician's fallacy," he says. "By which I mean, the statistician's job isn't to tell people what they want to know. As a scientist, I'm capable of deciding for myself what I want to know. And I don't want to know the probability that a theory is true. Or rather: I don't believe it's achievable. I'd like it like I'd like world peace. In theory, I'd like to get there. But knowing which theories are true is regrettably beyond us."

It's not that he denies that, say, the hypothesis that hammers fall faster than helium balloons is more plausible than the hypothesis that undergraduates are psychic for porn. "Popper would say it's not about plausibility or epistemology, but about having been severely tested. And the theory of gravitation has been more severely tested than the theory of precognition, so I'll build on the first and not on the second. It's not about belief and you can't quantify it. I assume it's true without assigning it a probability."

What I find interesting, talking to Lakens, is how far he agrees with Bayesians. In his fascinating online statistics course, he does a segment on Bayes early on, and makes it clear that you can't assess the chance that a hypothesis is true without a prior probability. He's also clear that a p-value of X means very different things depending on that prior probability. He cites Ioannidis, for instance, on why most published scientific findings are false, and points out that it's because most of the research was carried out into things that are a priori unlikely.

In fact, he acknowledges that he does something implicitly Bayesian. When he chooses what to research, he chooses a hypothesis that he thinks is a priori likely to be true. "Am I going to study precognition? No. In that sense I'm implicitly using Bayesian decision-making. My prior is low that it's going to yield something of value, so I'm not going to do it. As a scientist—as a human being—I use it to pick topics. But my evaluation of the data is not based on my priors."

Instead, he says, once you've got the data, you let the p-values stand for themselves. "You don't believe in the Higgs boson because you had a prior, you saw data and you updated it," he says. "They did two five-sigma tests [each equivalent to p-values of about 0.0000003]. Either the result is true, or we live in the only one out of 11 million universes where we got this level of data by fluke."

Instead of relying on priors, he says, we should be working to get better data—raising our standard for what counts as statistically significant, say. "If I need to impress people, I'd lower the error rates," he says. "Strongly lower the alpha level [the technical term for 'level at which you count something as statistically significant']. Then if I find something it's highly unlikely to be a fluke." That's easy in physics when you have a particle accelerator, or in genetics with huge genome-wide association studies, but he says it can be done in social sciences as well. "The Many Labs studies [Brian Nosek's coalition for replicating studies] reaches the same threshold. Meta-analyses [aggregations of earlier studies] use five-sigma thresholds. Or sometimes people do it implicitly—like the Food and Drug Administration says you have the normal alpha level, but you have to do it twice: so one 5 percent might be a fluke, but a second is 5 percent of 5 percent, so it's very severely tested."

POPPER AND HIS SWANS

Daniël Lakens invoked Karl Popper in his rejection of Bayesianism, so perhaps I ought to say what Popper argued.

Back in the eighteenth century, David Hume raised *the problem of induction*. All of our scientific reasoning, he said, is based on an assumption that the future will be like the past. If I drop a hammer and a helium balloon 1,000 times, and the hammer hits the ground first each time, we assume that lets us predict that it will do so the 1,001st time.

But the only reason that we think the future is like the past is because, in the past, it always has been. "[All] our experimental conclusions proceed upon the supposition that the future will be conformable to the past," Hume wrote in *An Enquiry Concerning Human Understanding*.[25] Using the same evidence to demonstrate that the future *will* be like the past "must be evidently going in a circle, and taking that for granted, which is the very point in question." Perhaps on the 1,001st time, we will release the hammer and it will fly due magnetic north, or it will hover while rotating around its long axis, or it will turn into a hummingbird, while the helium balloon crashes solidly to the floor.

Of course, we *do* use the past as a guide to the future, and "none but a fool or madman will ever pretend to dispute" the fact. But Hume struggled to understand how you could base that reasoning on solid philosophical grounds. He said that ultimately our expectation that the future should be like the past was due to "custom." "Perhaps we can push our enquiries no farther, or pretend to give the cause of this cause," he wrote, "but must rest contented with it as the ultimate principle, which we can assign, of all our conclusions from experience."[26] Hume believed that we just have to take it as an axiom, unprovable, that "the future will resemble the past." And its rational basis notwithstanding, empiricism—experience, observing the past and drawing conclusions about the likely future—was still reliable.

Understandably, philosophers weren't very happy leaving it at that, and the problem of induction has been a niggling thorn in their side for the last 250 years. It's a particular pain for philosophers of science—and philosophically inclined scientists—who want to say that when we do some study showing that some drug cures a disease or that uranium-238 decays into lead-206, we are not just saying that something happened *once*, but that it will continue to happen in the future. We want to be able to use that drug to cure other people, or use that uranium to fuel power plants or blow up cities.

Some philosophers—notably Paul Feyerabend—argued that this meant that all science was irrational, and that there was therefore no reason to think any one scientific theory better than any other. (When asked why, in that case, he tended to fly in airplanes rather than on brooms, he replied, "Because I know how to use planes but don't know how to use brooms, and because I can't be bothered to learn."[27])

Karl Popper, the great Austrian-British philosopher of science, tried to sidestep it, arguing that science didn't rely on induction at all. He said that when scientists test theories, they don't confirm them—they just fail to falsify them. His famous example was that of a simple hypothesis, *All swans are white*. Imagine that you see a white swan. Does that prove that all swans are white? No, of course not. We could see another white swan; it still won't prove it. There is no number of white swans that you could see before you could say with certainty that *all swans are white*. That's simple Aristotelian logic. You can't infer a universal law from individual examples: the syllogism "This is a swan, this swan is white, ergo all swans are white" is not a valid one.

What you *can* say, though, is that if you see a swan that *isn't* white, the statement *all swans are white* cannot be true. The universal statement *all swans are white* denies the possibility of any black (or green, or multicolored) swans. If you see even one, then you have falsified the hypothesis *all swans are white*.

Popper thought that this was how science advances—not by confirming true scientific hypotheses, but by falsifying false ones. "I happen to believe that in fact we never draw inductive inferences, or make use of what are now called 'inductive procedures,'" he wrote. "Rather, we always discover regularities by the essentially different method of trial and error, of conjecture and refutation, or of learning from our mistakes."[28]

You might feel (and I would agree with you) that this doesn't seem to be the whole story. The theory of aerodynamics has not been falsified, and nor has the hypothesis that there is alien life on Europa. But I feel great confidence in the theory of aerodynamics—I will even fly for thousands of miles through the air in a metal tube, supported by nothing more than the pressure differential between the upper side of the wings and the lower side, because I trust the theory and its practical applications. I feel much *less* confidence in the aliens-on-Europa theory. It could well be true, but so far no one has gone to check, and I would not bet on it unless at highly favorable odds. Naively, at least, Popper's model suggests that these two hypotheses are equal in validity.

Popper would say, though, that there is a difference—one has been severely tested, and one has not. "We choose the theory which best holds its own in competition with other theories; the one which, by natural selection, proves itself the fittest to survive," he wrote. "This will be the one which not only has hitherto stood up to the severest tests, but the one which is also testable in the most rigorous way."[29] He called such a theory "corroborated."

I'm not a giant of modern philosophy like Popper is (although I got a good grade in my master's!), so I'm somewhat outgunned intellectually here. But I must admit that I think this is a bizarre position. The idea that a "severely tested" or "corroborated" theory isn't in some way *more likely to be true* than a theory that hasn't been is strange. If you (or Popper) were to place a bet on whether some outcome predicted by the theory of aerodynamics was true—say, that my Boeing 777 will successfully leave the ground when it reaches 165 mph on the runway—and another bet on whether or not there are alien fish on Europa, you (and I expect Popper) would be willing to bet at much lower odds on the plane thing. That's because you've seen much more evidence of the first than of the second. Saying that the plane thing "has been more severely tested" seems indistinguishable in all but semantics from "is more likely to be true."

I'm not alone in my distrust of Popper. "Popper! Come on!" says Eric-Jan Wagenmakers, a professor in the stats and methodology unit of the University of Amsterdam's Psychology Department. "Popper said some pretty strange things," he says. "None of our hypotheses are ever true, but some are more easy to reject than others? In which case, why would you try to falsify them in the first place?"

Unlike his fellow Dutchman Lakens, Wagenmakers is—by his own admission—"a militant Bayesian. Not as militant as Aubrey [Clayton] but still, pretty militant." So he suggests the use of—unsurprisingly—Bayesian methods instead.

The problem that Popper's falsification model has is that it doesn't actually help. Most scientific hypotheses are not straightforwardly falsifiable by a single counterexample. If I were to hypothesize that "acetaminophen cures headaches," I'm not arguing that acetaminophen cures *every single* headache—if I gave you acetaminophen and your headache didn't go away, that wouldn't disprove the hypothesis. In fact, I'm not even claiming that acetaminophen cures *most* headaches. I'm only claiming that, statistically, you are more likely to recover quickly from a headache if you take acetaminophen than if you don't (or if you take a placebo).

This is (as I understand it) where Popper's and Fisher's approaches come together. Neither man would ever say that a hypothesis has been confirmed here, only that it has not been rejected. Fisher would say that you tentatively behave as though a hypothesis is true if your p-value is below 0.05; Popper would say that you can call it "corroborated" if it has been tested a lot and not found wanting. They just don't put numbers on it.

For Bayesians like Wagenmakers, though, this is just hiding from reality, a position forced on frequentists by their upstream decision to reject Bayesian priors. "If they admit that it makes sense to encode prior knowledge as numbers," he says, "they'd have no choice but to become Bayesian. So they take refuge in intuitive priors—they intuitively reason in a way that makes sense, but only informally." That is, he says, what Lakens is doing when he says he is Bayesian in his choice of topics to research, or what Popper is doing when he says that some hypotheses have been "more severely tested" and so you should build on them, while others have not.

In fact, late in his career, Popper did attempt to quantify "corroboration."[30] But the resulting equation ends up as functionally equivalent to the "relative belief ratio"—a Bayesian measure.[31]

Wagenmakers also strongly disputes Lakens's suggestion that, actually, researchers aren't interested in the inferential probability question. To reiterate: frequentist statistics answers the question "How likely are we to see this data, given a hypothesis?" What I have been suggesting researchers want to know—although Lakens and other frequentists deny it—is the inverse question: "How likely is this hypothesis to be true, given the data we've seen?"

"The questions you can address with frequentist statistics are of no interest to researchers!" says Wagenmakers. "It gets the conditioning wrong. We don't *want* to know how surprising the data is if the null hypothesis is true; we want to know the plausibility of the null hypothesis, now that we've seen the data. Ultimately, fundamentally, that's the question."

He and some collaborators tested this hypothesis by asking lots of authors of papers published in the journal *Nature Human Behaviour* about their beliefs. "We asked them about the claim in the main title of the paper," he said. "'Men enjoy eating apples more than pears,' or whatever. And we asked the authors what the plausibility of the main claim was before they saw the data and after. *Every* researcher, which is rare in science, said the data had made the claim more plausible than it was before.[32] But those questions are beyond the realm of frequentist statistics!"

If you read scientists' writing, they do tend to talk about the probability of their hypotheses. Here's Einstein, for instance: "I knew that the constancy of the velocity of light was something quite independent of the relativity postulate and I weighted which was the more probable."[33] And again: "[Abraham and Bucherer's] theories should be ascribed a rather small probability because their basic postulates concerning the mass of the moving electron are not made plausible by theoretical systems which encompass wider complexes of phenomena."[34] Scientists clearly *do* think in terms of the probability that their hypotheses are correct, not just whether or not they've been falsified. *Instinctively*, at least, scientists think like Bayesians.

BAYES AND THE REPLICATION CRISIS

Here's the basic advantage that Bayesians have over frequentists: they don't have to leave data on the table. Jens Koed Madsen, the LSE psychologist I mentioned who uses both frequentist and Bayesian methods as the mood takes him, puts it like this: "Frequentists have this weird thing where they have to jettison everything else. It makes it incredibly volatile." That is, every time they do a new study, in theory at least, all the information from the previous studies just gets forgotten. The hammer-falls-faster-than-the-helium-balloon hypothesis starts from scratch just as the undergraduates-are-psychic-for-porn hypothesis does. That means that your beliefs about the world can be blown around very easily, like leaves on the breeze. And *that* means that, as Madsen puts it, "it's easy to find a significant effect. You can fudge around with p-values because you always assume this is the first study to look at it."

For instance—I'm taking the numbers in this example from Daniël Lakens's marvelous online statistical inference course on Coursera, which I really cannot recommend highly enough—imagine you're collecting some data. Say, are people with red hair more likely to eat soup? You know the background rate of soup-eating in the population, so you just get two hundred redheads, ask them, "Do you eat soup?," and record the results.

Now *as it happens*, gingers are no more likely to eat soup than the rest of us. (Let us stipulate, for the sake of argument, that I haven't actually tested this hypothesis.) But as we've seen, the nature of the $p = 0.05$ cutoff means that if I were to run the experiment twenty times, I'd expect, on average, to see a significant result once. That's what a p-value of 0.05 means, remember: a result so extreme that you'd only expect to see it once in every twenty experiments, if there was no effect.

Imagine you do your experiment. But, after asking your first ten redheads about their soup-eating habits, you stop and have a look at your data. And if your p-value is below 0.05, you say, "Well, looks like I've found a significant result!," and you run off to *Nature* to get your paper published. If it's not, you carry on, and check after each new redhead whether the situation has changed. This doesn't seem like it should matter very much. Isn't it just saving time? And, in fact, in a lot of cases, it could save lives—if your vaccine trial, say, is showing strong early results, then it's important to know, so you can start getting it into arms rather than waiting months for more results.

But amazingly—to me, at least—this relatively innocuous bit of peeking early at your results changes your chance of getting a statistically significant result enormously. If the null hypothesis is true, and there's nothing there to see, then without peeking early, you'd see a $p \leq 0.05$ (that symbol means "equal to or less than") result one time in twenty; with peeking early, that jumps to about one in two.

Here's a graph, plotted in some statistics software using a script of Lakens's,[35] showing an example of how the p-value jumps around if you check your data after every new entry:

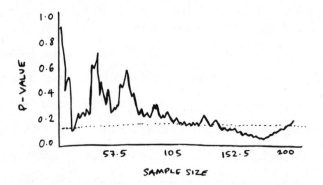

The dotted gray line is at $p = 0.05$. If the solid black line drops below it, then your data is (for that brief moment) statistically significant at the 0.05 level. In this example, it drops below it twice—a researcher could have stopped at either of those moments and declared a finding. Even though we know (because we know how the software has been programmed!) that there is no real effect here.

I ran this script a few times, and the black line always wobbled about dramatically. About half of the time, it dropped below the dotted gray one in those first two hundred observations. If you were an unscrupulous researcher, or even if you were just a naive one, you could very easily find apparently statistically significant results in noisy data when there's nothing really there, just by checking your data a few times before you originally intended to.

If you were a Bayesian, though, this wouldn't be a problem. You've already got the data from your priors—whatever they are—so each new data point coming in moves your opinion much less. And, of course, each new result forms part of the new prior for your next bit of information.

Dennis Lindley, the Bayesian primarch of Las Fuentes, argued that "the experimenter can go on sampling until he has reached the significance level α, and yet the fact that he did so is irrelevant to a Bayesian." [36] Others—including Ward Edwards,[37] a twentieth-century American psychologist and Bayesian, and Eric-Jan Wagenmakers[38]—go further, and say that optional stopping is actually a *good* idea for Bayesian analysis. One 2014 paper [39] ran a simulation a bit like the one I ran above (but rather more sophisticated) and showed that if you stop collecting data whenever your posterior probability or your Bayes factor—which I'll explain later, but for now you can think of it as p-values for Bayesians—drops below a certain point, then you'll still get (on average) the same probabilities as you would if you'd waited until all your results were in, as planned. But you'll get them quicker, so you can get your drug to the market or your newly discovered subatomic particle in the papers or whatever sooner and can move on to the next thing.

Another, more technical advantage that Bayesian techniques have over frequentist ones is that they don't just reject or accept the null hypothesis—they don't just say yes or no to a hypothesis, but give degrees of belief to a range of possible realities. That's important because—in reality—there is no such thing as a null hypothesis. Or rather, when looking at populations of human beings, the null hypothesis is always, ultimately, false.

Imagine you do some study looking at the difference between two sections of society. Let's say it's "Do redheads like soup?" again. If you look at two hundred redheads and two hundred people with other hair colors, you'll find some small difference just by chance, and under the frequentist framework you must decide whether that difference is large enough to reject the null.

But if you looked at *every single redhead and non-redhead in the country*, there would be some difference. Even if it was literally that there was one more soup-liking redhead per million. You'd be able to find *something*. So if you get a large enough sample size, you'll definitely be able to reject the null hypothesis. And it will be a real result. The University of Chicago psychologist David Bakan wrote in 1968:

> Some years ago, the author had occasion to run a number of tests of significance on a battery of tests collected on about 60,000 subjects from all over the United States. Every test came out significant. Dividing the cards by such arbitrary criteria as east versus west of the Mississippi River, Maine versus the rest of the country, North versus South, etc., all produced significant differences in means. In some instances, the differences in the sample means were quite small, but nonetheless, the p values were all very low.[40]

The great psychologist Paul Meehl said something similar—he once surveyed fifty-seven thousand Minnesota high school students, asking them about their religion, their leisure habits, their birth order, their number of siblings, their plans after high school, and dozens of other subjects. In total, the different responses could be mixed together in

990 different ways: Are students who like cooking more likely to be only children, are students from Baptist families more likely to join political clubs at school?—that sort of thing. Meehl pointed out that when he sliced up the data, 92 percent of those possible combinations came back with statistically significant correlations.[41] And these are real differences, with (presumably) some real, if multifaceted and complex, causes behind them.

Similarly, if you took thirty thousand redheads and thirty thousand non-redheads, you'd find some difference in their soup-eating habits. It would almost certainly be statistically significant. It would be real, not a false positive. But it's not clear that it would tell you something important—it might be a tiny correlation, or it might disappear when you look at a slightly different group of redheads.

The nature of frequentist statistics requires that you either reject the null or you don't. Either there's a real effect, or there isn't. And so, if you get a big enough sample size, you'll definitely find something. A Bayesian, instead, can make an estimate of the size of the effect and give a probability distribution.*

A probability distribution is a graph of the things that *could* happen. If you were to do a graph of the possible outcomes of a roll of a single six-sided die, then you'd have a graph with six equally tall bars on it, one marked 1, one marked 2, and so on up to 6, for each of the possible

---

*I should note that frequentists aren't idiots, and they have thought of all this stuff. You can estimate effect size within the frequentist framework, you can do "equivalence testing" that lets you determine whether an apparent effect is big enough to pay attention to, and, as we've discussed before, rejecting or accepting the null hypothesis isn't final. But it's somewhat ad hoc, not built into the system as it is with Bayes.

outcomes. Each bar would have a probability of one-sixth, or 16.7 percent, or 0.167, adding up to one, because you've got to roll *something*. (Technically there should be a seventh bar, "something weird happens," which includes outcomes such as "the die lands cocked ambiguously between two numbers" and "the die turns into a mongoose," but for the sake of argument let's say we're sure it'll end up on one number or another.) That chart would look like this:

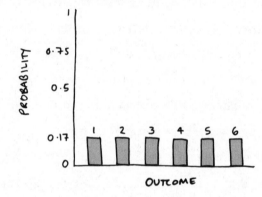

If you roll two dice, you get a different graph. It looks like the normal distribution—there are six ways to roll a seven (1 + 6, 2 + 5, 3 + 4, 4 + 3, 5 + 2, 6 + 1), but only one way to roll a two or a twelve (1 + 1, 6 + 6). So the probability of each outcome is different:

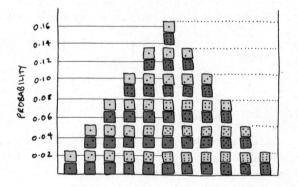

And if we were measuring some continuous variable, like height or weight, rather than a discrete one like the score on a pair of dice, you'd have a graph of a continuous curve, which might be a normal distribution or some other shape, depending on what you were measuring. But just as with the others, the area under the curve would add up to one, and if you wanted to measure the probability of seeing some result, say "the percentage of men between 172 cm and 178 cm tall," you'd look and see how much of the graph fell between those two points on the x-axis:

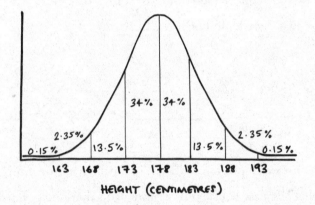

This isn't just a Bayesian thing, I should say. That distribution would make perfect sense to Jacob Bernoulli.

What makes it Bayesian is that, first, you can use a probability distribution to represent your subjective beliefs, based on the information you have, about your best guess concerning some topic. And second, you can update it with new information, to give you a new probability distribution.

Here's what that means. First, you have your prior distribution. That is, before seeing your data, you have a best estimate of the size of whatever the effect is. Let's use the "Do redheads like soup?" question again, and say that your best estimate is that redheads eat exactly the same amount of soup as everyone else. But you're not sure. The real effect could be that they eat slightly more, or slightly less, or, less likely but possible, that they eat loads more, or loads less. You can be pretty much certain (probability ≈ 1) that they eat *some amount of soup* between "zero soup" and "all the soup in the world."

The more confident you are that the real answer is close to some particular value, the more probability you place close to that value. So if you're very confident, your probability distribution is tall and skinny; if you're unsure, it'll be low and flattish.

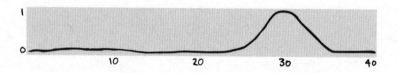

Then you examine some redheads' soup habits. You find, to your surprise, that they eat considerably more soup than the population average. The new data is distributed around an average. The new curve is called your "likelihood."*

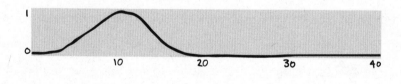

---

*This is an annoying thing statisticians have done, to take a word that in everyday language means exactly the same as "probability," and give it a technical meaning that is different in a subtle but important way, so everyone gets confused. Again. Like they did with "significance."

Then you multiply the posterior and likelihood together to make a curve that is the average of the two previous curves. That's your posterior distribution.

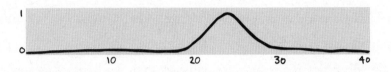

What the posterior looks like depends on how strong your prior was, and how good the new data is, in terms of sample size and effect size and so on. If your prior is very strong—your curve is really tall and narrow—and the new data is fairly weak, giving a likelihood curve that is low and wide, then the resulting curve will look more like the prior. If your prior is weak but you've got really good data, so your likelihood curve is tall and pointy, the new data will wash out the prior, and the posterior will be more like the likelihood.

But you don't have to say that you reject the null or accept the alternative hypothesis or anything. You just say, "I think the true value falls somewhere in this curve, with this probability for each point." If the posterior curve is tall and pointy compared to the prior, then you have something noteworthy and worth following up.

Bayesians like Aubrey Clayton believe that this would mean that the old problem of chasing significant p-values would be alleviated. You can always find a statistically significant result if you chop your data up enough, or if you check your results repeatedly, or if you just get bigger and bigger samples until you can find tiny meaningless correlations. The nature of the scientific publishing industry is that, often, you can get your paper published if you do find one of those meaningless (or spurious) correlations; and the nature of academia is that you need to get papers published if you want to progress in your career. Removing the binary concept of statistical significance and replacing it with a smoothly analogue "How big is this effect and how likely is it to be real?" chart goes some way to avoiding some of those bad incentives and cliff-edge cutoffs.

To be clear: replacing frequentist analysis with Bayesian analysis would not magically solve the many problems that modern science faces, and a lot of its problems could be solved within a frequentist framework. (And I shouldn't overstate the problems of science too much, either—yes, there's a lot of garbage and bad incentives, but science is still the reason you live a longer, richer, healthier life than your great-grandparents did, and the reason you can talk to anyone in the world using a six-by-four-inch box in your pocket.)

Even Wagenmakers, the arch-Bayesian, agrees with that. "It's not a magic pill," he says. "There are some core principles that hold whatever system you use. Don't cherry-pick, be honest. Garbage in, garbage out." If scientists are hiding their unsurprising null results, or journals aren't publishing them, then the scientific literature fills up disproportionately with "surprising" but false ones. That means when you try to do a meta-analysis to assess the overall state of the scientific consensus, you'll get a false picture, and that's true whether you analyze the data with Bayesian or frequentist methods.

But one thing that a Bayesian approach would fix is the use of the $p = 0.05$ threshold. And that's important, because—even though scientists often assume, or pretend, or imply, that a $p = 0.05$ result should be taken as real—p = 0.05 is sometimes actually evidence *against* your hypothesis. In the next section, I'll explain why.

DENNIS LINDLEY'S PARADOX

"A one-in-twenty chance of seeing results at least as extreme as these" sounds like a fairly high bar. That's what p = 0.05 means, as we've been discussing, and that's the standard for declaring that you've found something. (Or for "rejecting the null hypothesis," anyway.) But a one-in-twenty threshold like that is, in fact, surprisingly uninformative. In some scenarios, getting a p-value of around 0.05 is evidence *against* your hypothesis.

I'll try to explain why. P-values look at how surprising the data is under one hypothesis. "But in Bayes, you compare two hypotheses," says Wagenmakers. "And data can be surprising under the null, but even more surprising under the alternative." This is called Lindley's paradox, after a 1957 paper by the aforementioned Dennis Lindley, "A Statistical Paradox."[42] But as Lindley himself noted, it was present in Harold Jeffreys's work twenty years earlier. And it's not, in fact, a paradox; it's just that if you ask different questions of the data, it will give you different answers.

The idea is that, if you performed some experiment a large number of times—say a hundred thousand—and if, in reality, there is no true effect, you would tend to see p-values randomly scattered around. I'm going to use an example from Daniël Laken's Coursera course again:

Let's imagine you selected a group of people and measured their IQ. You know the population IQ is 100 (by definition). Let's imagine that there's no effect—that the population you're sampling from also has a mean IQ of 100. If you made a graph of the p-values you found in your one hundred thousand experiments, it would look like this:

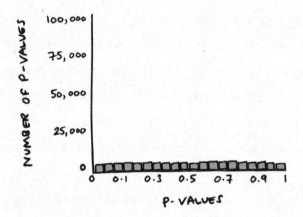

Sometimes you'd see extreme results, sometimes you'd see less extreme ones—that is, sometimes, just by fluke, you'd take a sample that had unusually high or low IQs, but they wouldn't be representative of the population; other times, you'd get a more representative sample. A p-value of 0.05 or below should happen one time in twenty. So should a p-value of between 0.05 and 0.1, and between 0.1 and 0.15, and so on. And you can be more specific: a p-value of between 0.04 and 0.05 should only happen one time in one hundred, etc.

But now imagine there *was* a real effect. Say you've gone and measured the IQ somewhere full of really clever people, and the average IQ there is 107. If you've got a decent sample size, and there's a real effect, you are *very* likely to get a really low p-value. Now your p-values will cluster enormously very close to zero. Instead of the flat graph above, you'll see something like this:

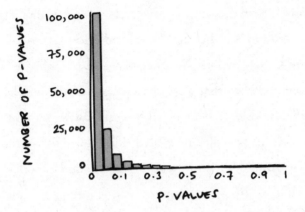

Very few p-values *get as high* as 0.04. So given those two hypotheses—either the population you're measuring has a normal 100 mean IQ, or it has a much higher 107 IQ—you're much more likely to see a p-value of 0.04 under the null. It's surprising, yes—but much *more* surprising under your alternative hypothesis.

Of course, those aren't your only two hypotheses. The real average IQ could be 94, or 110, or anything else. But if you don't have good reasons to favor any particular hypothesis very much—if your priors are wide and diffuse—then a just-as-about-significant result could well be better evidence for the null than against it.

This doesn't mean that the whole concept of p-values is bad. Kevin McConway, a professor emeritus of statistics at the Open University (who himself is Bayesian-sympathetic without being dogmatic about it) says it's just that the two frameworks are answering different questions, as we've been saying all along. A p-value of 0.05 tells you, correctly, that your data is surprising, given the null hypothesis. But it doesn't tell you anything about how likely the null hypothesis is, given your data. That's just not within its gift.

The trouble is that, too often, researchers assume that a statistically significant result means that they have good evidence for—or even confirmation of—their hypothesis. And that's just not what it means.

We should be clear that a large percentage of the problem is that a $p = 0.05$ threshold is almost laughably weak. If I were to put forward a hypothesis that my new dice are loaded, and I rolled two of them and got two sixes, that would be comfortably enough to declare statistical significance ($p = 0.028$). I play tabletop war games quite a lot, and I promise you that rolling lots of sixes happens all the time. (For my opponents.)

"From a Bayesian perspective, .05 is very weak evidence," says Wagenmakers. "The bar is set ridiculously low."

Of course, if the problem is that the evidentiary standard is too low, one obvious answer would be to raise the bar. That is precisely what Lakens was proposing earlier—reducing the alpha level (that is, the p-value required to declare statistical significance). In 2017, a group of scientists published a paper in *Nature Human Behaviour* [43] calling for the standard level of statistical significance to be redrawn at 0.005— that is, one in two hundred. It wouldn't eradicate Lindley's paradox, but it would make the number of situations in which it was relevant much smaller.

But the problem would remain that p-values don't actually tell you what the probability is that some hypothesis is correct. They would be stronger evidence, sure, and if you followed Fisher or Popper you would perhaps be more confident in your "corroboration" or your willingness to act as though the null had been rejected. But you still wouldn't be able to put a number on your belief, and if you agree with Wagenmakers, Clayton, et al., that that's what science is trying to do, then you need Bayesianism, and if you need Bayesianism, you need priors.

But . . . where do you get them from?

FINDING YOUR PRIORS

Here's a simple thought experiment, to see if you are, instinctively, a Bayesian or a frequentist. (I've borrowed this example from Cassie Kozyrkov, a decision scientist at Google.)[44] Flip a coin. Catch the coin, but don't look at it. (Hopefully, you've done it in the stylish way where you flip it with your thumb off your curled index finger, and as you've caught it in the same hand, you've slapped it down onto the back of your other hand. But however you've done it, you now have a flipped coin, covered by your hand.) What's the probability that the coin is heads? Come up with your answer before you carry on. OK. Now get someone else to flip a coin. They catch it. They *look* at it. What's the probability that the coin is heads now?

If you answered "50 percent" (or "0.5," if you're being proper about it), then you're thinking in a Bayesian way. Probability for you is about your own subjective beliefs, and about the information available to you. The coin could be heads, or it could be tails, and you have no reason to believe either is more likely than the other, so the probability is 50 percent. It doesn't make any difference *to you* when the other person looks at the coin—*for them*, the probability is now 100 percent or 0 percent. But for you, you have gained no new information, so the probability is still 50 percent.

If you answered, "Either 0 or 100 percent," or possibly "What the hell are you talking about?," then you're thinking like a frequentist. There is a right answer, a fact of the matter—either the coin landed heads up or it didn't. It doesn't make sense to talk about the "probability" of a thing that's already happened. (It also didn't make any difference when the other person looked at it. They now *know* the true answer, and you don't, but still, the true answer is there, and the probability is one or zero.)

This is what we mean when we say that Bayesianism is subjective. Probability and statistics should be seen as the assessment and measurement of uncertainty—we don't know whether X or Y will happen, but we can try to say *how likely* they are, given what we know about the world—and what I know about the world, and therefore how likely I think they are, might be rather different to what you know.

But there are two kinds of uncertainty. *Aleatory* uncertainty is the uncertainty in an unknowable future—*aleatory* coming from the Latin word *alea* meaning "a die." ("*Iacta alea est,*" as Caesar said, according to Suetonius, as he crossed the Rubicon and marched on Rome—"The die is cast," meaning that the consequences of his decision were coming, whatever they were, and they were unknowable.[45])

So before you flip a coin, there's aleatory uncertainty over whether you'll get heads or tails. When you get on a plane, there's aleatory uncertainty (though not very much) over whether it will land safely. When you drop your toast, in the second or so before it lands, there's aleatory uncertainty over whether or not it will land butter side up or down.

But then there's *epistemic* uncertainty, from *epistēmē*, the Greek word meaning "knowledge." That's what Cassie Kozyrkov was demonstrating above. If you flip a coin, then you *catch* it, but don't look at it—then there's no aleatory uncertainty. The result is there, it's happened, that's it. Still, though. You don't have any new information. As far as you're concerned, the question is no closer to being resolved than it was before.

Similarly, if someone you know is on a plane, there's epistemic uncertainty over whether it's landed or crashed (although I slightly regret this choice of example because planes are quite incredibly safe). Once you've dropped your toast, but before you look under the table to retrieve it, there's epistemic uncertainty over whether it's smeared butter over your kitchen floor.

Questions of real-world facts have epistemic uncertainty. What's the population of Switzerland? I don't know, but I'd say probably about 10 million. I'm 90 percent sure it isn't lower than 4 million or higher than 30 million. If I were to draw a probability distribution, that would be my prior: a curve with a peak at 10 million, with only 5 percent of my probability mass above 30 million and another 5 percent below 4 million. It'd look something like this:

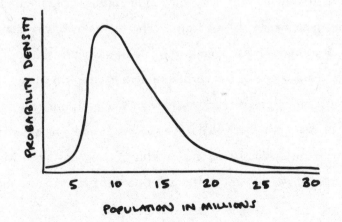

(I've just looked it up and the population of Switzerland on January 1, 2022, according to Eurostat, was 8,736,510.[46] Now my probability density function shrinks to a needle-sharp point centered around that figure.) With questions with discrete answers, like "What is the state capital of Georgia?," you can put probabilities on all the different answers: 60 percent on "Atlanta," 35 percent on "Tbilisi," that sort of thing.

All of which is fine. Daniël Lakens, certainly, would agree with us so far. And when we come back to the idea of Bayes' theorem as decision theory, and the more informal use of Bayes as a model for predicting the future and changing our mind and things like that, then this is really the only way to go about it. To the extent that the brain is a Bayesian machine—another idea we'll come back to later—this is pretty much what it's doing, when it predicts the world around you and updates it with new information from your senses.

But *in science*, how does it work? Where do you get these priors *from*? Are you allowed to just pluck them out of the air? Do you go, "I dunno, I reckon it's about 40 percent likely that this vaccine prevents COVID," and work from there, or is there some more sophisticated method?

Well, there are several, of course. The most obvious is just to say that we don't know. If you literally have no idea whether the population of Switzerland is one or the entire population of Earth, then you put equal probability mass on each possible answer, and your prior is as flat as your prior for whether you'll roll a six or a one on a six-sided die. "A uniform flat prior doesn't assume anything," says Jens Koed Madsen. "Frequentists have that prior, although they don't want to talk about it.

I say to my hardcore frequentist colleagues that their subjective prior is 0.5, and where does that come from? It's not explicitly specified in your theorem, but it's implied."

There are problems with uniform priors. The main one is Boole's objection, which we heard in chapter 1: that a uniform prior in one sense gives you a nonuniform one in another. The example we discussed was, predictably enough, an urn filled with balls, either black or white. If you have a flat prior on the total number of black balls in the urn, then any given mix of black and white balls is equally likely. (If there are only four balls in the urn, you have three possibilities—two black, one black, and zero black—and they're all equally likely.)

But if you assume that *each ball* is equally likely to be black or white—a flat prior on the probability of drawing a white or black each time—then your prior probability favors (very strongly, if there are lots of balls) a roughly fifty-fifty mix in the urn as a whole.

Harold Jeffreys suggested a route around this problem for many cases—a prior probability distribution that looks like a U, with the probability mass concentrated heavily at the extremes. (That is, you start out thinking that whatever you're looking for either happens almost every time, or happens almost never.)

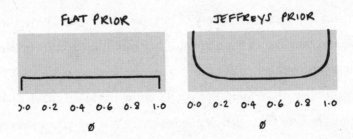

163

As with the flat prior, it's non-informative—that is, if you get some new data, your posterior probability will look pretty much like that data, without any real input from your prior. But it's also less vulnerable (though not completely so) to giving weird paradoxes where total ignorance in one sense gives you very strong prior beliefs in another.

Those sorts of priors are useful when you don't know anything at all. But total ignorance is unusual. You might not know very much about Switzerland, but you'd probably be pretty confident that more than ten people live there, and fewer than a billion. So most of the time you have some prior information that you'd want to include.

"In my previous work," says Madsen, "we were looking at fishing behavior in Indonesia. We wanted to understand the behavior of these fishermen, so we spoke to them, and to local NGOs and experts. And it seems kind of silly to me to say, 'I couldn't possibly integrate these experts into my prior, it's not data.' Imagine if you had some experts who looked at your model and said [about some aspect of it], 'Never, ever, has that happened,' then it seems arbitrary to set that prior to 0.5 just because it's the data thing to do."

But that means making subjective decisions about your prior. If you think it's more likely than not that Indonesian fishermen will use trawl nets rather than long lines, or if you think it's more likely that they'll catch tuna than octopus, then you have to say, "And I think they are 1.5 times as likely" or whatever. And that will influence the results you present when your actual data comes in.

Doesn't that undermine the whole point of doing the actual data collection? No, says Eric-Jan Wagenmakers. "You can check the robustness of your conclusions by checking different prior distributions." So you might try asking whether your conclusions stand up if you think tuna are 1.7 times as likely to be caught, or 2.4 or 1.3.

"Usually," he says, "it doesn't really matter as long as it's reasonable. And most people agree if it's reasonable or not. And usually, because the data will just tell you a clear story, it doesn't matter so much." If it *does* matter a lot, then your data probably isn't very good.

Andy Grieve, the pharma statistician, tells a similar story. "For very early studies, or internal ones, we'd use subjective information," he says. "You can elicit information from experts, for instance. We'd use that in internal decision-making.

"But it's pretty unlikely that you'd be allowed to do that in a submission to a regulatory authority, so in bigger trials we'd use what information we have on the drug, or similar drugs, from historical data."

Lakens, the frequentist, is very skeptical of all this, and in fact expressed doubt that anyone used the results of previous experiments to form the priors of the next ones. "Did you manage to find a scientist who ever used Bayes' theorem to actually update their prior in practice?" he asked me. "As in, they published a paper in 2018, and then actually used the result from the 2018 paper as a prior, collected data, and reported an updated quantified posterior belief? Anyone, ever, updating their belief a single time, in a published paper?"

But Wagenmakers disagrees. "Of course we use the posteriors!" he says. "If you didn't, you'd be willingly throwing away data. Money is on the line, in industry, so of course you don't."

Grieve, who did work in industry, says that's what his pharmaceutical research did all the time. It's just more efficient, he says. Normally, scientists pool all the studies on a subject and do a "meta-analysis"—

they use all the data from all of the studies, combining their p-values and effect sizes and so on to create a consensus. For Bayesians, though, that's just part of the daily work. You incorporate all the studies that were done before. "It allows us to leverage all the data we've collected in the past," Grieve says. "It incorporates its own meta-analysis. The current standard way of creating a prior distribution from existing data, in fact, is called a meta-analytic prior."

It's just a fact that Bayesian procedures make more efficient use of the data you have. "If you're not using a proper, informative prior, you're leaving money on the table," said one Bayesian, the US epidemiologist Robert Weiss.[47] If there is data, information, that you could use and which you're choosing not to—your eventual conclusions will be less certain than they otherwise would have been. There might be good reasons not to use the existing data, but not using it will make your use of any new data less efficient.

One problem is that you could, in theory, skew your results by taking strange priors. For instance, if you were running a pharmaceutical trial, you'd have a treatment group, getting your drug, and a control group, getting a placebo or standard care. If you dishonestly (or incompetently) arranged your prior expectation of the effect in the control group, so it looked much worse than it should have been, then it would make the apparent effect in the treatment group look much better.

"That's the big concern," says Grieve. "If when you collect your data there's evidence that your control arm is very different from what you've seen before, from your historical data." It's not that pharma companies engage in that sort of bad behavior regularly, he says, "but there's always the odd bad egg." The way to avoid it, he says, is "a mix of distributions for the prior, which automatically down-weights the historical information if there's a big difference between the historical data and the current data."

It's not that finding a prior is a trivial or obvious task. There are choices to be made, and (even though one school of Bayesianism describes itself as "objective") they can be debated. If you disagree on how trustworthy a study is, or whether to include expert testimony, then you might disagree on your prior probability distribution.

But that doesn't mean people have to pluck priors out of the air. There are reasonable ways of finding them in different circumstances. And then, of course, if your data is any good, your priors will be rapidly washed away.

YOU'RE NOT BEATING A DEAD HORSE IF THE HORSE IS STILL WINNING THE RACE

There's a temptation when you're writing about controversial topics to be Deeply Wise, and to say, ah, yes, people are angry, but both sides have very fine people on them, very fine people on both sides. And to be fair, that's not necessarily the wrong attitude to take. The Bayesian-frequentist row is remarkably bitter—I had one person describe someone on the other side as "a car salesman, out to win souls for the cause." Someone else described another as "the Donald Trump of methodology." (Very fine people on both sides.)

As Andy Grieve said, perhaps some of it is for show. Even Daniël Lakens, whom Bayesians think of as the arch-frequentist, says that "often frequentist approaches are best, but sometimes you do have enough prior information to say we can use Bayesian statistics, and in those situations it has clear advantages. That's the nuanced position, but you're not going to write a book saying that."

Cassie Kozyrkov, the Google data scientist, in her blog post about whether you're a Bayesian or a frequentist, has a subheading. "So, which one is better?" and her answer is: "Wrong question! The right one to choose depends on how you want to approach your decision-making."[48]

She also points out, probably rightly, that during her graduate studies at Duke University—"which is to Bayesian statistics approximately what the Vatican is to Catholicism"—the loudest voices shouting about how great Bayesianism is weren't the professors but the students, mainly because the basic Bayesian ideas are easier to grasp.

Sophie Carr, the statistician who runs the consultancy firm called Bays, is surprisingly nondogmatic about it as well. "I talk about frequentist and Bayesian statistics like rugby," she says. The two codes of rugby—league and union—have subtly different rules, and fans of the two different disciplines are loudly adamant that their version is the best. (For non-Britons, I think it's fair to say that league is a more working-class game, played mainly in the north of England; union is played more in the south of England, and in Wales, Scotland, and Ireland, and in England, at least, is more middle-class.)

"Leeds Rhinos were my team, and they're league," says Carr. "Then I came south and played union for Bath." You can switch between the two, and neither is better than the other, although each has pros and cons. Her analogy with frequentism and Bayesianism is obvious.

I am tempted, therefore, to say, "*Of course* this debate is highly charged, lots of emotions running high, but both sides make good points!"

It's obviously true that frequentist methods are perfectly good in lots of scenarios—Lakens is right that it would have been pointless including prior probabilities on the search for the Higgs boson, for instance, when you're dealing with p-values that you'd only see one time in 11 million or something if there wasn't a Higgs to find. DNA sequencing in biology—genome-wide association studies looking at the entire length of the genome in hundreds of thousands of people, and comparing them with phenotypic outcomes like diseases, height, intelligence, whatever—might not need Bayes either.

It's also obviously true that taking a Bayesian approach wouldn't, on its own, solve the problems that science faces. If journals still preferentially publish novel, surprising results over unsurprising ones, and if academics are still operating on a publish-or-perish model and need to get papers into journals if they're going to succeed, then there will still be perverse incentives in academia. It might make some difference if the statistics are analyzed with Bayesian methods rather than frequentist statistics—you can't p-hack if you're not using p-values, so at the very least we'd have to come up with a new name for it—but it won't solve the issues. If scientists won't share their data or their code for others to check, it doesn't matter whether that data was analyzed using Bayes factors or not.

And—to continue the theme—it's *also* obviously true that you *can* solve, or at least ameliorate, a lot of these problems *within* a frequentist framework. Some academics I know advocate something called Registered Reports, in which journals agree to publish papers on the strength of their methods, before the data is collected, so then, whether the researchers find exciting, headline-worthy results or boring, null results, those results will go on to become part of the scientific record. Several relatively major journals have signed up for Registered Reports, and I think they're a good idea—they remove the incentive to slice the data until you get a positive result, and they remove the problem of publication bias. They represent a helpful move, whether papers are Bayesian or frequentist.

And, as mentioned several times, a lot of the main problems with frequentist models are that the $p = 0.05$ threshold is laughably weak, and you could ameliorate those problems significantly by moving to $p = 0.005$ or something. Yes, there'd be a lot of studies that wouldn't get published (or, ideally, would be published with the headline "We Looked but Didn't Find Anything").

Another idea would be simply getting rid of the "academic journals" model altogether. I spoke to another psychologist, Marcus Munafo at Bristol University, about this, and he thinks that essentially the idea of the academic journal being the repository of record—the place where the scientific record is kept, the official store of science—is outdated. "The idea of three-thousand-word articles published in a journal is three hundred years old," he told me. "Research is more complex now, and we have the technology to present it all and the moving parts."

In fact, there is an alternative model already in place—Alexandra Freeman of the Winton Centre for Risk and Evidence Communication at Cambridge University has launched a program called Octopus. It is a free repository for hypotheses, data, code, and methods. Freeman, a former journalist, told me, "When I moved from the media to academia, it struck me that academics are being given the exact same incentives as journalists—they're pushed toward telling good stories, instead of doing good science. Journals encourage people to have high-impact publications, which they define as having high readership, short and to the point, carrying a message. It acts directly against what you actually *want* in a primary research record—which is everything there, in detail, so people can follow it."

Instead of scientists doing research and then, when it is completed months or years later, spending another few months or years hawking it around to publishers in the form of a written piece, Octopus is "designed with a completely different incentive structure," she says. "You publish your hypothesis to Octopus, then you come up with a method to test that hypothesis, and you link that to it. Then anyone can carry out the protocol you've described." Then you publish the data on it, and anyone can analyze the data. Meanwhile, journals can carry on disseminating interesting work—"They can be essentially like *New Scientist* or *Scientific American*. And they can have a paywall if they like. But Octopus is where you share the actual research, for free."

I understand people who say something like "Why are we spending so much time arguing over Bayesian versus frequentist statistics? Our entire scientific publishing system is screwed, academics' incentives are to churn out pap rather than uncover truth; it seems silly to worry about whether they do that with Bayes or with p-values." And there's also a sense of weariness, I think, in academic circles. Are we *still* arguing about this? Surely we've got better things to worry about now? Haven't Bayesians made their point?

But I do want to nail my colors to the mast to some extent. For one thing, even though Bayesian methods are far more common and widely accepted than they were fifty years ago, the standard techniques for investigating a scientific question are still frequentist. "Go on Google Scholar," says Aubrey Clayton, "and search for 'p-value' or 'significance' or whatever. It's still the common language. There are tens or hundreds of thousands of articles per year. Maybe the tide is shifting, but the dominant mode is still frequentist." People get annoyed at Bayesians for banging on about it, he says, "like we're beating a dead horse, rehashing this debate about Bayesianism. But David Bakan had this great line, 'You're not beating a dead horse if the horse is still winning the race.'"

And Bayes does have advantages. For one thing, it definitely *does* solve, or ameliorate, some of the issues of the replication crisis. Going back to the Lindley paradox: under frequentist analysis, a statistically significant result can actually be evidence *against* your hypothesis. Because Bayes forces you to compare the likelihood of seeing a result between the two competing hypotheses, it's much harder to say, "And this just-about-significant result supports the headline claim!" when your analysis shows you that it doesn't.

In theory, at least, some of the more direct HARKing methods, such as optional stopping, are not a problem for Bayesians. And, of course, if you're doing research into some unlikely hypothesis, such as psychic abilities, then you have to choose a prior that reflects that unlikeliness, and as a result the strength of evidence that you need will be greater.

The other advantage is that you get to make use of all the data available to you. Yes, in cases like the Higgs boson you have so much data that your prior doesn't matter. But in, say, vaccine studies, where you're trying to see how many people catch the disease in your control group vs. your treatment group, it might take months or years to obtain enough data to get below a certain significance threshold. But if you're allowed to use data from your earlier trials, and include them as prior probabilities, it gets you there more quickly. Not using a good, informed prior is leaving money on the table, to return to the Robert Weiss quote.

But Eric-Jan Wagenmakers makes a point that I also agree with, which is that Bayesianism is *aesthetically* more pleasing. "There's something in Bayes," he says. "Everything is coherent; you don't have internal inconsistencies. In frequentism you can find all these anomalous cases, and people say it's an anomaly but only in this situation, but it always feels ugly.

"Fundamentally, it's a matter of elegance, of aesthetic."

There's a wider point too. *Outside* of science, this is just how decision theory works. "What are we trying to do with classical statistics?" asks Wagenmakers. "We're trying to make a decision between two hypotheses. How could we do that, with Bayes? We'd specify our utilities and our prior probabilities, compute our evidence, and take the decision that maximizes our subjective utility. In economics, we'd say that's the normative way of doing it.

"But the p-value is a really bastardized version of that. No priors, no utility—that's all implicit. How that can be accepted as good decision theory is beyond me. No one would use it in human decision theory."

That might not have made a lot of sense, since I haven't yet explained what *utility* is. But the idea of Bayes as the underlying system for all decision-making is what we're going to look at next.

CHAPTER THREE

Bayesian Decision Theory

ARISTOTLE AND GEORGE BOOLE

Many readers will probably be familiar with logical syllogisms. The classic: All men are mortal; Socrates is a man; therefore Socrates is mortal.

This is deductive reasoning. If you accept the two premises (All men are mortal; Socrates is a man) then you have to accept the conclusion (Socrates is mortal) on pain of contradiction. The syllogism is *valid*, which is not the same as being true: it just means that the conclusion follows from the premises, not necessarily that the premises are true. For instance, "Plants are good for you; tobacco is a plant; ergo tobacco is good for you" is a logically valid syllogism, but it's factually incorrect.

The idea of deductive reasoning usually gets credited to Aristotle.[1] The physicist and probability theorist E. T. Jaynes, who plays roughly the same role in the cult of Bayes as St. Paul does in Christianity, says that pretty much the whole of Aristotelian logic can be boiled down to "the repeated application of two strong syllogisms,"[2] to wit:

> If A is true, then B is true.
> A is true.
>
> —
>
> Therefore B is true.

And the opposite:

> If A is true, then B is true.
> B is false.
>
> —
>
> Therefore A is false.

You can replace A and B with any propositions you like. If burbles are wurbles, then Abraham Lincoln was the forty-fifth president of the United States; burbles *are* wurbles; therefore Abraham Lincoln was the forty-fifth president of the United States. If fish could fly, my grandmother would be a bicycle; my grandmother is not a bicycle; ergo fish cannot fly.

As before, these statements are valid—if you accept the premises, you must accept the conclusion—but not, necessarily, correct.

You can add various elements to it: "If both A and B are true, C is true; A and B are true; ergo C is true" is a more complicated form of the first syllogism. "If A is true, both B and C are true; C is not true; ergo A is not true" is a more complicated form of the second. But those are the fundamental actions.

In the nineteenth century, George Boole, the aforementioned scourge of the uninformative prior, introduced the use of algebra to codify the whole deductive reasoning thing. So A $^\vee$ B means "both A and B are true" (conjunction). A $^\wedge$ B means "at least one of A and B is true" (disjunction). ¬A means "A is not true" (negation). A $\rightarrow$ B means "A implies B," or "if A is true, then B is true" (implication).

Then there are a bunch of axioms. If A is true, ¬A cannot be true. If A $^\wedge$ B is true, B $^\wedge$ A is true. That sort of thing. From those relatively simple atoms, you can build the whole world of propositional logic.

Aristotelian (or Boolean) logic has a simple job: it spits out a truth value. At the end of a sequence of logical statements, it will end up saying either "A is true" or "A is not true." And even though the bits from which it is made are relatively simple, you can do some complicated things with it.

Very complicated, in fact. The Boolean algebra can be represented as logic gates. A logic gate is basically a simple computer chip with two inputs, and depending on whether those inputs are active, it sends an output.

Let's imagine the logic gate is wired up to some simple inputs: a light sensor and a microphone, say. The light sensor fires if it's above a certain light level; the microphone fires if it's above a certain decibel level. The gate's output is attached to an LED.

If you attach them to an AND gate, then the gate will give off a signal and light the LED if and only if both the light sensor and the microphone are firing—so if it's bright AND noisy.

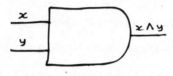

If you attach them to an OR gate, the LED will switch on if it's noisy OR bright (or both).

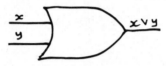

There's also a NOT gate, which fires if it's NOT receiving some input—so you could attach it to the light sensor, and it will keep the LED lit as long as it hasn't got any light on it.

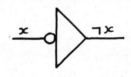

Those gates do exactly what the Boolean operators do. An AND gate (which in our example would say, "If [light] and [noise] are true, then [LED is on] is true") is the same as the logical conjunction, $^\vee$. It's a syllogism. An OR gate ("If [light] or [noise] is true, then [LED is on] is true") is the same as the logical disjunction, $^\vee$. A NOT gate ("If [light] is not true, then [LED is on] is true") is the same as the logical negation, $\neg$.

Those simple systems are enough to do all the calculations that a fully functioning digital processor can do (although you also need some memory to build a real, working computer). In fact you can do it even more simply, with a NOT AND (or NAND) gate, which always fires *unless* both its inputs are true—you can use NAND gates to build all the gates described above. For instance, an AND gate could be made by using a NAND gate with a split output that then goes into a second NAND gate. If both inputs to the first gate fire, then it won't fire; which means that the second one will.

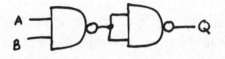

I am writing this on a computer whose CPUs could be entirely simulated using NAND gates. Propositional logic is powerful stuff. George Boole went so far as to describe its operations as "the laws of thought."[3]

But it's also limited. If we want to work out whether something is true or not, usually we can't work it out with logical certainty. We might want to say, "If it is Friday, my children will have fish for their school dinners; it is Friday; ergo my children will have fish for their school dinners." If we accepted the premises, believed them to be unarguably true, then we would be forced to accept the conclusion. But we can't be sure of the premises. Perhaps the school has run out of fish and they're having lasagne today. Perhaps we've got our days wrong and it's actually Thursday.

Or, to cite Jaynes: A policeman, late at night on a deserted street, hears a burglar alarm coming from the broken window of a jewelry store. A masked man emerges from the broken window carrying a bag; upon inspection, the policeman finds the bag to be full of gold and gems. The policeman assumes the man to be a thief, and most of us would agree that that's the likeliest explanation. But there's no logical certainty about it. "It might be, for example," says Jaynes, "that this gentleman was the owner of the jewelry store and he was coming home from a masquerade party, and didn't have the key with him. However, just as he walked by his store, a passing truck threw a stone through the window, and he was only protecting his own property." [4] It doesn't sound very likely, I admit, but you can't prove it's not true.

And yet most of the time, this is the sort of reasoning we have to make do with. We can't do the full logical syllogism. We must make do, as Jaynes says, with "weaker" ones. Instead of "If A is true, B is true; B is false; therefore A is false," we have to make do with things like:

If A is true, B is true.
B is true.

—

Therefore A is more plausible.

If it will rain by 10 a.m., then there will be clouds in the sky before 10 a.m. It's 9:45 a.m. and there are clouds in the sky. Therefore, it's more plausible that it will rain at 10 a.m. This seems a decent analogue of how we actually think and reason.

It's not just that we use new information—we also base our reaction to that information on our prior experience. The brain "makes use of old information as well as the specific new data of the problem," says

Jaynes; "in deciding what to do we try to recall our past experience with clouds and rain, and what the weatherman predicted last night." If, Jaynes says, every night the policeman went out, he saw a masked man emerging from the broken window of the jewelry shop, and every time it turned out to be the lawful owner, then pretty soon that policeman would stop paying attention. "Thus, in our reasoning we depend very much on *prior information* to help us in evaluating the degree of plausibility in a new problem," says Jaynes. "This reasoning process goes on unconsciously, almost instantaneously, and we conceal how complicated it really is by calling it *common sense*."

So we have prior information; we get new information; we combine the two together to form a revised picture of the world. Doesn't that sound familiar? Almost . . . Bayesian?

Yes, it does. And what Jaynes (and Jeffreys, and modern decision theorists) would say is that Bayes' theorem is, indeed, how reasoning works—not just our common sense, but all decision-making under uncertainty. And in fact they would argue that Aristotle's and Boole's idea of logic is just a slimmed-down, special-case version of Bayesian reasoning—one in which the probabilities are set, implausibly, to one or zero, absolute certainty. Bayesian logic, on the other hand, lets us deal with all the shades of gray in between.

BAYES AS THE HEART OF DECISION-MAKING

Classical logic deals entirely in ones and zeros—which is fine, if we are using it to prove logical statements, or to run central processing units.

But if we want to talk about probabilities—making decisions under uncertainty—we need the in-between numbers.

More than that, we need—or, at least, we *want*—a mathematical framework for moving between those numbers, for changing our beliefs about something. It will not come as a surprise to learn that the appropriate framework for doing so is Bayes' theorem.

As David Manheim, an academic and superforecaster (don't worry about what that is if you don't know; I will explain it later), told me: "Scientists can be frequentists or Bayesian; decision theorists cannot. There is no way to do decision theory with frequentist math."

I'm going to take a simple example from a guy called Eliezer Yudkowsky, who deserves a book of his own,* but who will only be described here as the author of *Rationality: from AI to Zombies*, where this little thought experiment appears.[5] Imagine there's a state lottery where you live. The lottery draws six numbers out of a possible seventy; you need to have all six numbers to win the jackpot.

There are, therefore, 131,115,985 possible combinations of numbers, and if you have one ticket, you have a 1-in-131,115,985 chance of winning.

That's your prior probability, and it's not huge. But imagine now that you also plan to cheat. You have a box that beeps when you give it the correct, winning combination of numbers! Now, admittedly, you have a lot of numbers to put into it, but in theory, if you keep hammering away, one number a second for four years, you'll find the right number.

---

* I've sort of written one—*The Rationalist's Guide to the Galaxy*, available in all or at least some good bookshops and probably several indifferent bookshops too!

There's a drawback, though. Your box also beeps at random, one time in four, even when the combination is wrong.

So you run a combination through the box. It beeps! What do you do? Do you run off and buy a ticket? After all, there's only a 25 percent chance of a beep on any given wrong number!

Hang on though: you're forgetting your priors. In Bayesian terms, your data—a beeping box—is four times more likely under the hypothesis "This is the correct combination" than under the hypothesis "This is the wrong combination." That is your *likelihood ratio*, which we've talked about before: 4:1.

But if you were to run your beep-box machine on all 131,115,985 possible combinations, it would beep roughly 32,778,996 times. Only one of those combinations is actually correct. So you take your prior probability (1:131,115,985) and your likelihood ratio (4:1). Then you multiply 1:131,115,985 by 4:1 and you get a new posterior probability—4:131,115,985, or one in 32,778,996.

To think of it another way, you have your prior probability distribution, with every single possible combination having exactly 1/131,115,985 of the available probability mass. You then do your box thing. If it doesn't beep on a number, you can be sure it isn't the winning combination, so you can reduce the amount of probability mass you put on that number to zero.

(One thing, which we'll come back to. I have assumed that I could be certain, if the box didn't beep on a given combination, that that combination was not the winning one: I assigned it zero probability. That's cheating, really. I should have assigned it some negligible but non-zero amount—perhaps the box is faulty! Or perhaps I missed the beep! But I shall do it as Yudkowsky did, in order to keep the math simple.)

The box beeps (on average) on one combination in four. So you push your probability mass onto those combinations. Each of those numbers now has 1/32,778,996 of the probability mass.

If you run the box over those numbers again, and remember the false positives are random, it will once again beep for the correct combination, but it will also beep for (on average) 8,194,749 wrong combinations. You'd have to run a ticket through the box fourteen times, with it beeping each time, in order for that ticket to be *likely* the right one.

These are just *the rules*. You can't, as Yudkowsky puts it, "stop on the first combination that gets beeps [ten times in a row], saying 'But the odds of that happening for a losing combination are a million to one! I'll just ignore those ivory-tower Bayesian rules and stop here.'"[6] If you did, you would still have less than a 1 percent chance that you were holding the right ticket.

Now, it might well be a good bet to buy that ticket—that depends not just on the probability of winning the prize, but on the value of that prize if you do win it; decision theory needs to talk about the *utilities* of the various outcomes as well as the probabilities of them, which we'll come back to—but the fact is it's still not very likely to be the right ticket.

There's an idea in thermodynamics called the Carnot engine, after the nineteenth-century French mechanical engineer Nicolas Carnot. It's an idealized heat engine: the most efficient engine theoretically possible using a heat-exchange system. Any real engine—a steam engine, an internal combustion engine—will be less efficient, because heat will be lost to the environment, so it will do less work for a given amount of energy than a Carnot engine would. But as your engines become more efficient, they approach the efficiency of Carnot's model.

Bayes' theorem is to decision theory what the Carnot engine is to thermodynamics. The analogy is Yudkowsky's again,[7] and it's a solid one. You can't run a real car on a Carnot engine. You can't build one. It's an idealized, imaginary model, to which any real engine can only be an approximation. But the real engine is only working insofar as it's approximating the Carnot engine, and it's not working insofar as it's not.

Similarly, you'll rarely be able to apply Bayes' theorem perfectly to real-world situations. You can't *perfectly* determine the prior probabilities of, say, Russia invading Ukraine, or the local shop having run out of pink grapefruit squash. And you can't *perfectly* determine the strength of the evidence that you get to update those priors—if satellite images show a buildup of Russian armored divisions in Crimea, how much should you update? If the Safeway website says pink grapefruit squash is in stock, how much should you trust it? Your estimates of all these probabilities will be approximations.

But when you make decisions—when anyone does, or when any agent or decision-making process does—you do so by approximating Bayes' theorem. A decision made under uncertainty is good insofar as it approximates Bayes' theorem and bad insofar as it leaves Bayes behind.

What E. T. Jaynes demonstrated in his posthumously published work *Probability Theory: The Logic of Science* was that by using Bayes' theorem, as we have above, we can do all the things we can with Aristotle's logic, and more. "Aristotelian deductive logic is the limiting form of our rules for plausible reasoning," says Jaynes.[8]

That is, if you use only the probabilities one and zero, you can do all the same logical moves within a Bayesian decision framework as you can in Boolean or Aristotelian logic. "If men are mortal with probability one, and Socrates is a man with probability one, then Socrates is mortal with probability one." Or, to revisit our generalized versions:

If the probability of A is 1, then the probability of B is 1.
The probability of A is 1.

—

Therefore the probability of B is 1.

And the opposite:

If the probability of A is 1, then the probability of B is 1.
The probability of B is 0.

—

Therefore the probability of A is 0.

All the operations we looked at in Aristotelian/Boolean logic can be carried out in this mode, if you limit yourself to using ones and zeros. The rule that A and not-A cannot be true at the same time is the same as saying, "The probability of either A being true or A not being true is one." The "logical conjunction," the AND gate, can be represented as "$p(A \wedge B) = p(C)$," where A and B are the inputs and C is the output.

(Or, to put it in plain language, the probability that both A and B are true equals the probability that C is true; the probability that the light sensor and the microphone are both signaling is the same as the probability that the LED will be switched on.)

But what Bayesian probability theory can *also* do, says Jaynes, is give us something like common sense. Remember, most of the time—almost all the time, really—we can't do deductive, Aristotelian logic; we can't say, "If A then B; A, therefore B." We have to say, "If A then B; B, therefore A is more plausible."

For instance: If it has rained overnight, the pavements will be wet in the morning. The pavements are wet. Therefore, it's more plausible that it's rained. It's not definitely true; perhaps your sprinklers were on. But the hypothesis *it has rained* is more plausible, given the evidence, *the pavements are wet*. And the great thing about Bayes' theorem is that you don't just have to say "it's more plausible": you can put numbers on exactly how much more plausible it is.

So let's say that when it rains, you see wet pavements 80 percent of the time. When it doesn't rain, you still sometimes see wet pavements—say, your sprinklers come on 20 percent of the time. You're four times more likely to see wet pavements under the hypothesis *it has rained* than under the hypothesis *it has not rained*. That's your likelihood ratio, and that tells you how much to update your beliefs—how much more plausible the "rain" hypothesis is, given the "wet pavements" evidence.

That's not all we want to know, though. We want to know *how probable it is that it rained*. And for that (once again) we need prior probabilities.

Let's say that at this time of year, it rains 33 percent of evenings.

That's your prior probability. Now let's imagine that you see wet pavements one morning. What's the probability that it was raining last night?

Imagine you watch the pavements for one hundred mornings. On average, it will have rained thirty-three times, and not rained sixty-seven times.

On the sixty-seven mornings when it hadn't rained, you'll have seen wet pavements 20 percent of the time—13.4 mornings, on average—and dry pavements the remaining 53.6 mornings.

On the thirty-three mornings when it *had* rained, you'll have seen wet pavements 80 percent of the time—26.4 times—and dry pavements 6.6 times.

So if you have wet pavements in the morning, it means last night was either one of the 13.4 non-rainy nights with wet pavements or one of the 26.4 rainy nights with wet pavements. Which means the probability of it having rained last night is now 26.4 divided by (26.4 + 13.4) or 39.8, or 0.66.

As with Aristotelian logic, this is just what you get if you accept the premises. If you agree that rainy nights happen 33 percent of the time, and that wet pavements happen 80 percent of the time after rainy nights but only 20 percent of the time after dry nights, then those numbers ineluctably lead you to agree that, if the pavements are wet, there is a 66 percent chance that it rained last night. It's as unavoidable as the "All men are mortal; Socrates is a man; ergo Socrates is a mortal" syllogism. Except you can use it not merely to say, "This statement is true or not true, given these premises," but also "This hypothesis is this probable, given this evidence."

Of course, in reality you won't always have the precise numbers, and there's more than just one piece of relevant information. If we could really just read out a couple of simple numbers off some universal database and plug them into a one-line equation to work out how probable everything is, then predicting the future would be easy. *Actually* the probability of rain would depend on a million factors—the time of year, the air pressure, cloud cover, temperature, humidity, the number of butterflies to have recently flapped their wings in Brazil—and it would be computationally impossible to do the sums, even if you could keep track of every single one. But *if you could*, you could simply run Bayes' rule over all of them, and give yourself the precise probability of rain tomorrow.

CROMWELL'S RULE

As a great Bayesian thinker put it: *I beseech you, in the bowels of Christ, think it possible that you may be mistaken.*

The thinker was, of course, the Lord Protector Oliver Cromwell, writing to the General Assembly of the Church of Scotland in 1650, ahead of the Battle of Dunbar.[9] Cromwell had the misfortune to die more than forty years before Thomas Bayes was born, but nonetheless he has given his name to an important rule of Bayesian decision theory, Cromwell's rule. Named by Dennis Lindley, the rule says that you should never assign anything, other than a logically necessary truth such as "2 + 2 = 4," a probability of one or zero. That is: you should never be certain.

One should always, says Lindley, "leave a little probability for the moon being made of green cheese; it can be as small as 1 in a million, but have it there since otherwise an army of astronauts returning with samples of the said cheese will leave you unmoved." [10]

Here's another reason why. Let's go back to our rain-and-wet-pavements example. Imagine that my house is in the Mc-Murdo Dry Valleys in Antarctica. (The schools aren't great and you can't get a good coffee, but the house prices are *very* reasonable.) There hasn't been any rain for 2 million years. So my prior probability of rain on any given night is about 1 in 700 million.

I notice wet pavements. I know that the likelihood of seeing wet pavements is four times greater under the hypothesis "it rained last night" than under the hypothesis "it didn't rain last night," we established that last time, but the prior probability of rain is so low that my posterior probability is still vanishingly tiny, about 1 in 170 million, or 0.000000006.

But then I walk down the street, and I notice that the pavements outside my neighbor's house are also wet. Using my posterior probability as my new prior probability and doing the same sums again, I end up with a new probability of 0.000000024, or one in 42 million. Still, it's amazingly unlikely that it rained. My sprinklers and my neighbor's sprinklers must have gone off at the same time.

Then I walk past another house. Then another. Both pavements are wet. My probability has gone up to 0.000000384, about 1 in 2.5 million. Still, the overwhelmingly more likely hypothesis is "All our sprinklers happen to have gone off at the same time," but the "rain" theory is starting to become less insane than it was.

By the time you've walked past sixteen more houses, and noticed the wet pavements outside all of them, it's about 70 percent likely that it rained last night. It doesn't take *all* that much evidence to shift you away from even very, very strong priors.

But now imagine your prior was zero. You plug that into your equation, and your posterior probability would be—zero. Whatever evidence you find, zero times anything equals zero. You could walk past a thousand houses, all with wet pavements, and you would never assign the faintest probability to the idea that it might have rained.

ON DESCRIBING PROBABILITIES AS ODDS

If we think about probabilities in terms of odds, rather than percentages or a number between zero and one, the reason to avoid ones and zeros becomes clearer. You get odds by taking the probability and dividing it by 1 minus the probability. If something is probability 0.9, then you take 0.9 and divide it by 1 minus 0.9, or 0.1. So it's 0.9/0.1 = 9. Your odds are, therefore, 9:1. If it's 0.5, then your odds are 0.5/0.5, or 1, so they're 1:1.

When you use probabilities, a probability of one looks much the same as a probability of 0.9 or 0.5—it's just one more number. But when you use odds, it's clearly very different. A probability of 0.999999 in odds is 999999:1, but a probability of 1 equals infinity to 1. Infinity isn't a real number, and you can't use it in sums like a real number. (To steal a line from Yudkowsky again: "People sometimes say something like, '5 + infinity = infinity,' because if you start at 5 and keep counting up without ever stopping, you'll get higher and higher numbers without limit. But it doesn't follow from this that 'infinity − infinity = 5.'"[11])

Odds have another advantage, which is that they show the real differences between seemingly similar probabilities. The difference between 0.99 and 0.999 in probability looks small—smaller than the difference between 0.5 and 0.51—but in odds, it's the difference between 99:1 and 999:1.

You should never assign anything a probability of zero or one. To be clear, that doesn't mean you can't *act as if* things are impossible. It's not impossible that all the atoms in a statue's arm happen to move back and forth at the same time, making the statue wave at me. It's not impossible that I throw a hundred thousand heads in a row on a fair coin. But both occurrences are so unlikely that they will never happen in the lifetime of the universe, or several trillion universes. Very, very small probabilities are very, very small—you don't have to think, "So you're telling me there's a chance?!" when you hear "There's a one-in-a-quadrillion chance." But Cromwell was right: *I beseech you, in the bowels of Christ, think it possible*—if not necessarily likely—*that you may be mistaken.*

CONSERVATION OF EXPECTED EVIDENCE

There are some fun things that fall out of the Bayesian decision system. One is that you can't go looking for new evidence to support your theory; it is impossible, because any evidence you find must (in expectation) be as likely to reduce your belief as increase it, and if you *don't* find evidence, that is in fact evidence against your hypothesis.

Say that I think that a politician is bad. I expect that politician only to do bad things, like torture puppies. I *want* to bolster my belief that they are bad and evil. So I go and look at their reelection website, searching for any pro-puppy-torture policy positions, which I am confident I will find. Either I see them, or I don't. How would either of those outcomes change my belief?

The temptation is to think that if I find something, it will increase my confidence, but if I don't find anything, it will have no effect. But that's not how it works. If some piece of evidence would shift your belief by some amount, then the *absence* of that evidence must shift your belief in the opposite direction, and by an amount proportionate to how strongly you expected the evidence.

Say that you think it's 95 percent probable that you'll see puppy-torture advocacy on the politician's website if they're bad. That means you think it's only 5 percent probable that you won't.

Of course, if the politician is *not* bad, you'll be less likely to see the puppy torture. Say that if they're not, then you'll only see it one time in ten.

Here's how it would play out. If you see a puppy-torture video, as you expected, then that would shift your beliefs somewhat—up from $p = 0.9$ to $p \approx 0.99$. But, *because* it's expected, it doesn't shift your views all that much.

But if you *don't* see the video—if you are *surprised*—then it must move your beliefs a long way. In this case, your strong expectation being confounded would lead to your belief in the politician being a dog-torturer crashing to just p ≈ 0.33, one in three.

Again, this is unavoidable. If some evidence is strongly expected, then it can't move your beliefs very much; it's already part of the model of the world that you've built. But if something really unexpected happens—or, in this case, if something expected *doesn't* happen—it should move your posterior belief significantly.

In fact, the two are inversely proportional—the more strongly you expect something, the less you are surprised (and so the less your posterior probability changes) when you find it, and the *more* you are surprised (and so the *more* your posterior probability changes) when you don't. And what that means is that, on average, your posterior probabilities should be exactly equal to your prior probabilities—if you'd expect to see some evidence nine times out of ten in a universe where your prior belief was true, then *not* seeing it should shift your beliefs by nine times as much as seeing it. If you expect to see it ninety-nine times out of a hundred, then not seeing it should shift your beliefs ninety-nine times as far as seeing it would have.

This also means that, pace the common saying, absence of evidence is, in fact, evidence of absence. If I don't believe in unicorns, then I don't expect to see any unicorns. Every second that goes by without my seeing one is some small, weak evidence in favor of my "unicorns do not exist" hypothesis, shifting my probability estimate a tiny bit toward one. But, of course, if I see a single unicorn, that would be utterly devastating to my hypothesis and my posterior probability would be way down.

If your reasoning doesn't work like this—and for a lot of us, it doesn't, especially on political questions, because we're prone to confirmation bias and groupthink—you are simply *not making good use of evidence*; you are not updating your beliefs in the best possible way. If, for instance, you strongly expected to see the video, and then didn't, and shrugged your shoulders and said, "Well, she's probably bad anyway," then you're going to make yourself more wrong than you need to be.

UTILITY, GAME THEORY, AND THE DUTCH BOOK

The point of Bayesian decision theory is to help make a decision. Or, more accurately, to describe the optimal way of making a decision, given uncertainty about the outcome.

So far we've only really talked about how people form beliefs, and how they change the probabilities they attach to those beliefs, given new evidence. A good Bayesian should multiply their priors by the likelihood and form a new posterior probability that is a mix of the two. That's the Bayesian epistemology, and—as we've seen—it might be practically impossible to gather all the evidence and to compute all the sums, at least within the lifetime of the universe, but it is the correct way of working out how much probability you should assign to a given hypothesis.

That's not the same, though, as saying that it tells you what to do in a given situation. For that, you don't just need beliefs and probabilities, you need to know how much you care about something. In decision theory, that's called *utility*.

The easiest way to think about utility is to pretend it's the same thing as money. Obviously it's not, but since we earn money through the use of our time and labor, and we have a limited amount of it that we have to ration out between the things that are most important to us, it's a pretty good proxy, and also allows you to do straightforward sums.

Probability and utility together make *expected value*. To understand what that means, let's go back to that imaginary lottery and the beeping box, the example I borrowed from Eliezer Yudkowsky a few pages ago.

Before your box beeps even once, the chance of any given lottery ticket being the winning one is 1 in 131,115,985. That's not a very good chance. But you don't yet know if it's a bad idea to buy a ticket. If the tickets cost £1 each, and the value of the jackpot is £150,000,000, then if you were to buy one ticket for every possible combination, you would be guaranteed to make money—£18,884,015, a tidy sum. Even if you couldn't buy all of them, it would be a good idea to buy as many as you can reasonably afford: each individual ticket is worth, on average, £1.14. That's simply the value of the jackpot divided by the chance that you'll win it. So at £1 each, every ticket sold is an expected net loss for the owners and an expected net gain for you, of 14p. That's the expected value of buying a ticket.

If the tickets cost £2 each, then, obviously, you'd be down 86p, on average, on each ticket you bought. If you ran your beeping machine over a ticket, though, and it beeped, then your posterior probability of the ticket being correct is a mere 1 in 32,778,996. Suddenly, each ticket is worth 150,000,000/32,778,996, or £4.58, so on average you'd be up £2.58 on each one.

This is the concept of utility theory, and, once again, it is mathematically inescapable: you can't avoid it, on pain of contradiction. This was what Frank Ramsey and Bruno de Finetti realized in the 1930s—if you don't obey the laws of utility theory, you become the victim of a "Dutch book." Here's what that means.

As Ramsey argued, we can represent our confidence in a belief with a bet. If you think something is 50 percent likely to happen, you should be willing to take a bet at even odds or better, because the expected value is positive. If you think something is 33 percent likely, you should be willing to take any bet at two-to-one odds or better, for the same reasons.

But that relies on your beliefs adding up to 100 percent, to probability 1. If they don't, you're going to end up paying out money, whatever happens. Say that I think it's 50 percent likely that it will rain tomorrow. I'd be willing to bet 50p, and if it rains, you give me £1 back, including my stake.

And I *also* think it's 60 percent likely that it *won't* rain tomorrow. So I'd be willing to bet 60p against a £1 payoff.

In that situation, you could offer me both bets: the "Dutch book." If I'm sincere about my beliefs, then I would be willing to accept them both. But the pair of bets together cost me £1.10, and the payout will be £1, whatever happens. I may as well hand over 10p before we even start (and, of course, my betting partner really ought to make a bigger bet). I have become strictly irrational, and you can just turn me into a pump for squeezing money out of.

As I say, probability theorists usually use money as an example, because it's nice and easy. Also, there's pretty good evidence that while people say "money can't buy you happiness," it sort of can—the GDP per capita of a country correlates pretty well with the quality of its citizens' lives.

Health economists and bioethicists do sums using "quality-adjusted life years" or QALYs—for instance, the National Institute for Health and Care Excellence (NICE) in the UK says that an intervention that saves one year of healthy life, one QALY, for less than £20,000 is generally considered cost-effective.[12] It also operates with the same model of expected value—a treatment that will extend the lives of 10 percent of patients by five years is better than a treatment that will extend the lives of 20 percent of patients by two years, because $5 \times 0.1 = 0.5$, while $2 \times 0.2 = 0.4$, and 0.5 is bigger than 0.4.

But both of these are only a proxy for the real thing. Utilitarian philosophers and economists think in terms of *utility*—that is, how much happiness we gain, or how much our preferences are fulfilled, by a given action.

John von Neumann, the great Hungarian-American polymath—inventor of game theory, pioneer of computing and quantum mechanics, owner of a Wikipedia page dedicated to "things named after John von Neumann"[13] that is three screens deep—was interested in this. Economists of his time wanted to describe a normative way to make decisions under uncertainty—that is, how best to choose between options, if you want to maximize expected well-being. He was trying to develop a model of economics that could select the decision that would make everyone happiest.

Economists in that era thought that this was fundamentally impossible, because it involved trading off incomparable things.[14] If I build a new out-of-town shopping mall, it provides me with money (which I want) and some people with convenience (which they want), but it also spoils the view of local people (which they don't want). How do we say, "This much convenience is worth this many uglified views?"

It's easy enough when you're just dealing with one person: von Neumann and his coauthor, Oskar Morgenstern, imagine Robinson Crusoe alone on his island. He might not be able to *fulfill* all his desires—if he wants a back rub then he's bang out of luck, let alone if he wants a penthouse apartment or a first-class airline ticket to Bali. But he is able to choose freely between all the desires he's able to fulfill. If it takes an hour to build a shelter out of banana leaves, and it takes an hour to build a fire and cook a yam for dinner, and he has one hour before it gets dark, then he can decide whether he values not being rained on or not being hungry more. He can simply list his desires in order of preference and fulfill as many of them as the tools and time available to him allow.

There's no mathematical problem here, and classical economics could handle it perfectly well—"Crusoe is given certain physical data (wants and commodities) and his task is to combine and apply them in such a fashion as to obtain a maximum resulting satisfaction," write von Neumann and Morgenstern. "[He] faces an ordinary maximum problem, the difficulties of which are of a purely technical and not conceptual nature."[15]

But once Man Friday arrives, you have a problem. Crusoe enjoys his carbs and doesn't mind sleeping under the stars, so he would choose food over shelter; Friday doesn't like yams and he gets cold easily, so he prefers shelter to food. Suddenly, if you want to maximize the *group* utility, you must trade off one person's desires against another's; they have conflicts of interest. The two people are trying to maximize different things.

Classical economics assumed that while you could rank people's preferences—Crusoe, (1) food, (2) shelter; Friday, (1) shelter, (2) food—you couldn't compare them. Even if Crusoe was *really, really hungry*, and Friday only a bit chilly, you couldn't make mathematical comparisons between them. But von Neumann realized that you could. Morgenstern would later describe the moment: "I recall vividly how Johnny rose from our table when we had set down the axioms and called out in astonishment: '*Ja hat denn das niemand gesehen?*' ('But didn't anyone see that?')."

Von Neumann set out a few simple axioms. A key one was that people's desires need to be *transitive*—that is, if they prefer A to B, and they prefer B to C, they must prefer A to C. If I like dogs more than cats, and cats more than gerbils, then I must like dogs more than gerbils.

If my desires are *intransitive* then, as with the Dutch book, I become a money pump. If I like dogs more than cats, and cats more than gerbils, and gerbils more than dogs, then if I own a gerbil, you can offer to sell me a cat, for £1 plus the gerbil. But then you prefer dogs to cats, so you offer to sell me a dog, for £1 plus the cat. And then, of course, you prefer gerbils to dogs, so you can sell me my original gerbil for another £1, take the dog back, and start the whole process again, £3 richer already and with a lucrative day ahead of you.

The preferences also need to be *continuous* and *monotonic*, which means that if one decision will give you a 50 percent chance of £10, you should be indifferent between it and a decision that gives you a

100 percent chance of £5, and that a 2 percent chance of an outcome is twice as good (or bad) as a 1 percent chance of an outcome. That means there are no sudden jumps in people's preferences—as an outcome becomes more or less likely, the expected value of that outcome goes up and down smoothly. Also, preferences should be *substitutable*—if you're indifferent between cake and jelly, then you shouldn't care if you're given a 10 percent chance of cake and a 90 percent chance of jelly, or a 90 percent chance of cake and a 10 percent chance of jelly.

Given those assumptions, von Neumann sketched out his utility theorem, in which people have preferences that can in principle be assigned a number (with units called "utils") and compared to one another. I might assign ten utils to a nice day in the park, one hundred to watching Liverpool win the soccer league, one thousand to hearing of the birth of my first niece.

This work of von Neumann's created the field of game theory. If I want to model how two or more people with different preferences interact—like our Crusoe and Friday above—then I need some model to compare those preferences. Once you've established that such a model can in principle exist, as von Neumann and Morgenstern did, you can start doing calculations and thought experiments.

For instance, to move away from one classic piece of British adventure literature to another, von Neumann imagined Sherlock Holmes fleeing Professor Moriarty. Holmes takes a train to the ferry at Dover and is spotted as he does by Moriarty on the platform. He knows Moriarty can catch the next, faster, train and beat him to Dover.

What should Holmes do? If he goes to Dover, Moriarty will be waiting for him. Instead, he should get off at Canterbury, the only intermediate station; he won't escape to the continent, and Moriarty will still be waiting for him at Dover, but at least he's evaded capture in the short term. But Moriarty knows that too—so perhaps Moriarty will get off the train at Canterbury and wait for him there. Should Holmes stay on until Dover? But then . . .

Von Neumann puts some numbers on the scenarios. Moriarty gets 100 utils if he catches and kills Holmes, either at Dover or Canterbury. He gets 0 (a draw) if he goes to Dover but Holmes gets off at Canterbury, meaning that the hunt goes on. He gets -50 if he gets off at Canterbury and Holmes rolls on to Dover and then to France.

HOLMES / MORIARTY	DOVER	CANTERBURY
DOVER	100	0
CANTERBURY	-50	100

What should Moriarty do? If he gets off at Dover, his average payoff is fifty: $(100 + 0)/2$. If he gets off at Canterbury, his average payoff is twenty-five: $(100 - 50)/2$.

So he should get off at Dover. But if it's obvious that he should do that, then Holmes will predict it, and will get off at Canterbury, and Moriarty's payoff will be zero.

The answer is that Moriarty should be unpredictable—if the game were played many times, he should go through to Dover three times out of every five, and get off at Canterbury two out of every five. That maximizes his expected utility, at forty utils each time. Meanwhile, Holmes should do the opposite, and get off at Canterbury three-fifths of the time. (In the book, Moriarty *does* go all the way to Dover, while Holmes and Watson get off at Canterbury and watch his train go by.)

In reality, of course, you can't know exactly what the expected utility of any decision is, just as you can't entirely compute the Bayesian probabilities of every outcome, because you don't have the information or the computing power. But if humans were perfect reasoning machines with full access to our own underlying preferences, we could work out the math, using a combination of Bayes and utility theorem. That is, roughly speaking, what modern artificial intelligence does, in a much more explicit way.

OCCAM PRIORS

The thing with Bayesianism is that you have to have priors. Throughout its history, that's been a sticking point—Where can you get them from? How much of a problem is it that they seem to be subjective?

Sometimes you can get your prior from easily available statistics about the world; for a cancer diagnosis test, the prior probability can be the background rate of that cancer in the population of people like the patient. But sometimes you can't be as precise as that.

In deciding between possible hypotheses, one way to establish your priors is to look at *which is more complex*. Things that are more complex are less likely to arise by chance—so all else being equal, presented with two possible explanations, one simple and one complex, your priors should favor the simple one.

There's a name for that—Occam's razor. It's named after a fourteenth-century Franciscan monk called William of Ockham, who lived in Ockham, Surrey.*

But how do we decide what the simplest explanation is? When we look at the world, often the explanations for things seem very complicated indeed. There's a saying, which Eliezer Yudkowsky ascribes to Robert Heinlein,[16] although it may be a misattribution. The simplest explanation, possibly according to Heinlein, is always "The lady down the street is a witch; she did it." If you're explaining why someone got ill, "a witch did it" does seem simpler than, for instance, "Billions of self-replicating particles got into your body, and started taking over the machinery of your cells to make copies of themselves, and the combination of that and your body's own attempts to fight the particles off are what made you ill."

Similarly, "The thunder god was angry" seems simpler than the equations of electrodynamics that physicists use to explain lightning. Certainly, the latter would take a lot longer to explain than the former. We *understand* gods (or we understand people, and assume gods are like people). We *understand* anger. Most of us don't understand calculus.

But decision theorists have a more formal definition of simplicity. The fact that something can be described in a short English sentence doesn't necessarily tell us very much about how simple that thing is—I can say "the human brain" in four syllables, but the brain itself is the most complex thing in the universe.

---

* So really it should be called William's razor. You don't call Lawrence of Arabia "Arabia" or Jesus of Nazareth "Nazareth."

Instead, decision theorists use something called *minimum message length*. (I could also describe it as "Solomonoff induction" or "Kolmogorov complexity." The three are subtly different, but at heart they're equivalent.) What minimum message length asks is: What is the shortest computer program I could write that would describe a given output?

Let's start with something simpler than the creation of the universe. I'm going to take my example from the Czech mathematician/computer scientist Michal Koucký, of Charles University in Prague.[17] Imagine three eleven-digit strings of numbers, he says. They are:

1. 33333333333
2. 31415926535
3. 84354279521

Are any of these random? If you wanted to write a program that would carry on those sequences to a million digits, how short could that program be?

You're not allowed to just say, "Print random numbers," by the way. A random number generator is equally likely to produce any one of those strings, at a probability of $p \approx 1/10^{11}$, which is to say very unlikely. You want to know—is there a deterministic process that could produce them so that you could predict what the next number in the sequence could be?

The first one is pretty simple. It would be a million digits of the number 3. You could write that in four lines of BASIC:

```
10 N = 1000000
20 FOR I = 1 TO N
30 PRINT 3;
40 NEXT I
```

The others are more complex. They appear fully random; Koucký says that a statistician looking at them would say that they pass statistical tests for randomness.

But actually it's very easy to predict what the twelfth digit of the second string would be, because the first eleven are the first eleven digits of pi. We can just look up the next digit, or, if we think that's cheating, we can work it out. Archimedes established a simple method of getting ever better approximations of pi before 200 BC, using regular polygons. Either way, the answer is eight. I could write out Archimedes's algorithm, or any one of dozens of alternatives, and if I plugged one of them into a computer it would (eventually) predict the string of numbers out as far as we wanted. You can compress an infinite series of digits into a few characters.

But the third string is truly random. If you wanted to describe it to the millionth digit, you'd have to write it out to a million digits. There is no shortcut that would let you do it any faster; you cannot compress it at all. The minimum message length of any given output is *how short your description can be*.

We're trying to decide the prior probability of various hypotheses, not write strings of numbers. But we can flip the idea around, and say that if we see a string of numbers, what's the most likely algorithm that produced it? Again, for the three strings of digits we looked at, a truly random number generator could have made any of them, with equal probability—there's about a one-in-10^{11} chance of it producing any of them. That would be the simplest explanation. But if you saw that it had generated 31415926535, you'd be unimpressed with the hypothesis "The numbers are random," and you'd think there was some slightly more complicated algorithm that better fit the data, such as "Print the digits of pi in order." Even though a random number generator would be just as likely to produce that string as any other, there's another hypothesis that would be much more likely to produce that particular string, so you're happy to accept a bit more complexity in order to have a hypothesis that more confidently predicts that data.

On the other hand, if you saw 84354279521, as far as you would know there's no particular pattern to the data. So you would have to take a big hit on complexity—the algorithm would have to say, "First print this digit, then this digit, then this one . . ." in order to explain it. So the hypothesis that this is just the product of a random number generator, no more or less likely than any other string of numbers, seems more plausible. The trade-off is between the complexity of the algorithm and how confidently it would predict the output.

So how do you decide how much to trade off between the two? Say you want to explain a series of coin tosses. You see, say, HTHHTT, and you have to choose between several different possible algorithms that could produce it.

The simplest is a program that says, "The coin is fair, and gives heads or tails at random." That would be a maximally simple algorithm and a very straightforward program to write. But it would assign equal probability to that series of results as to any other—it would say your chance of seeing that or any string of results would be one in sixty-four.

Alternatively, you could hypothesize that the program said, "The coin will come up H, then T, then H, then H, then T, then T." It would assign a 100 percent probability to that outcome—it would fit your data perfectly—but it would be much more complicated.

If you only care about how simple your algorithm is, you'll always say, "This coin is fair," even if the string of coins is HTHTHTHTHT or HTHHTTHHHTTT. And if you only care about how well it fits the data, you'll say every coin is fixed. But if you care about both, how do you decide how much weight to put on each?

The way to think about it is as information. A single "bit" of information—a binary one or zero, a yes-or-no question—is enough to divide a search space in half. Imagine a game in which you are observing someone trying to find a door with a prize behind it. There are one hundred doors, and you know the correct one, but they don't. The only way of communicating with them is with a light switch, on or off.

Before you start, the probability of the prize being behind any given door is p = 0.01. Your playing partner wants to get that probability up. They can say, "Turn the light on if the right door is numbered between one and fifty." You turn the light on. Now they know that it's behind one of the first fifty doors, so they can assign p = 0.02 to all the remaining doors. They've halved their search space, and pushed twice as much probability mass onto each remaining option.

That's how much you should trade off between complexity and good fit. If an extra bit of information in your program doesn't allow you to halve the search space, then it's not paying its way. It's not compressing the data—you're just shifting it into the program, rather than the data.

So in choosing between two or more hypotheses, you should (in theory) be able to look at which is the more complex, and—all else being equal—assign higher prior probability to the one that would be simpler to write as a computer program, and with each extra bit of information in the program reducing its probability by half. There are other ways of producing priors, but minimizing complexity like this is a key one.

This is, by the way, remarkably close to how modern AI systems do in fact work, when making a decision under uncertainty. Paul Crowley, a cryptographer at Google, told me that in the most basic forms of AI, "if you understand Bayes, then it really seems incredibly Bayesian." A modern neural network AI has lots of nodes, like neurons in a brain, and the way it learns is by strengthening and weakening the links between those nodes—giving them higher or lower "weights." "You penalize it for having a really complicated set of weights," says Crowley. "An answer that involves a simpler set of weights has a better score. Forcing it to choose simpler hypotheses over complex ones is precisely a Bayesian thing; it's an Occam prior." Doing the math in an explicitly Bayesian way is computationally expensive, so most modern AIs use "easier systems that are much lower compute and almost as good," says Crowley, but Bayesianism is the underlying mechanism.

HYPERPRIORS

You may remember that George Boole had an objection to Bayes. Say you have an unknown distribution of white and black balls in an urn. What's your prior? That each ball is equally likely to be black or white? Or that any combination of black and white balls is equally likely?

As we saw, they're very different things. If you've got two balls in the urn, and any combination is equally likely, then your three options—two black balls, one of each, or two white balls—each have a probability of ⅓. But if each ball is equally likely to be black or white, then all black or all white only have a probability of ¼, and one of each has a probability of ½.

If you do it with more balls in the urn, it's even more obvious. If you assume ignorance over the *total distribution*, the chance of seeing zero black balls out of a hundred is one in one hundred and one. But if you assume that *each ball* is equally likely to be black or white, the chance of seeing zero blacks is about one in a million quadrillion.

That's a problem, because it means you can't be truly ignorant. If you say you're ignorant over the total distribution of balls in the urn, then you are claiming some knowledge about the chance that the next ball will be white. If you're claiming ignorance over whether the next ball will be white, you're claiming some knowledge about the distribution of the balls in the urn.

You can get around this by thinking about *hyperpriors*. That is, you're not only uncertain about some parameter—the number of black balls in the urn, say—but, on a higher level, what parameter you should be using. Perhaps you should be looking at the probability of a given ball being black, instead. The higher-level parameter—which parameter should I use?—is your *hyperparameter*, and your prior beliefs over what hyperparameter to use are your *hyperprior*.

In a way, it's uncertainty over *what world you're in*. For instance, imagine you're a very simple Bayesian AI that's playing hide-and-seek. Your opponent is either hiding behind a tree or hiding behind a wall; you start out with a prior probability that they're equally likely. You play a thousand games, and in eight hundred of them your opponent hides behind the wall. So, in the traditional Bayesian way, you update your prior probability. You now start each game with an 80 percent estimate that your opponent is hiding behind the wall.

But then something changes. Over the next hundred games, your opponent turns up behind the tree eighty times.

A very simple Bayesian learning model might simply include that data in with the rest of it, giving you about a 75 percent chance. Or it could be a bit more sophisticated and weight more recent data more heavily, so that there's a higher chance.

Or it could recognize that the world has changed. It could build a new model, one in which the opponent has a preference for the tree over the wall. It could recognize that the world has two states, and be ready to switch its model between the two, looking out for evidence that predictions based on one model are failing to come true. If you see your opponent hiding behind the tree several times, you increase the probability that the world has changed, and that you should be using the behind-the-tree model.

That's a hyperprior—a higher-level prediction about the shape of the world, one that constrains and informs lower-level ones. Just as with the normal prior, you assign probabilities to how likely it is you're in one world or another.

MULTIPLE HYPOTHESES

Imagine you meet someone who says he is a psychic.[18] How much credence do you give his claim? As Oliver Cromwell (and Dennis Lindley) would tell you, it would be wrong to give it precisely zero. *I beseech you, in the bowels of Christ, think it possible that you may be mistaken.*

But on the other hand, it's not very likely, is it, whatever Daryl Bem might say.

Say you meet someone. Call him the Mysterious Barry. He tells you he can read your mind. If you write down some numbers between one and ten, he'll guess them. How many numbers would he have to guess correctly before you'd believe him?

Each correct guess has only a one-in-ten chance of happening by pure chance. So if, even after two correct guesses, you would still think it vanishingly unlikely that the Mysterious Barry is a real psychic, then presumably your prior probability is well below one in one hundred that his claims are real. Perhaps it would take ten correct guesses before you'd start to think it was a realistic possibility, which would suggest that you think it's around the one-in-ten-billion mark.

But Jaynes points something out. In that situation, even ten correct guesses—even a thousand correct guesses—probably wouldn't move you to thinking that psychic powers are real.

In the early 1940s, a British parapsychologist called Samuel Soal claimed that he had discovered evidence of psychic powers.[19] Two subjects in a card-guessing game got the answer right more often than chance would suggest—one scored 2,980 out of 20,000, when the expected chance result would have been 2,308; the other scored 9,410, when chance predicted 7,420. That second result is twenty-five standard deviations from the mean, which implies that you should not expect to see it by chance, even if you were repeating the experiment every second for the entire lifetime of the universe.

And yet even knowing that, I suspect that you are unconvinced that the experimental subject was really psychic. If there were really only two possible explanations for the data—"pure chance" or "psychic powers"—then, yes, any remotely plausible prior probability you had for the existence of telepathy would be washed out by this extraordinary level of evidence. But that's not the case.

There's another possibility—that it's not pure chance, but that it's not psychic powers either: that there's some other explanation for why the subject came up with the correct answers more often than you might expect. It could be fraud, it could be sloppy experimental design, it could be that it's all a big practical joke.

Since your prior probability of psychic powers being real is so extraordinarily small—somewhere around the one-in-ten-billion mark, we said earlier, somewhat arbitrarily—any one of these alternative hypotheses is almost certainly vastly more plausible. And any evidence that could support the "psychic powers" hypothesis presumably also supports any one of those.

So if to begin with your prior belief in psychic powers was one hundred times smaller than your prior belief in the possibility that a given scientific paper is fraudulent, then at the end that will still be the case, no matter how much evidence you've seen. Short of some experimental design that all but rules out the possibility of fraud, no amount of evidence could ever make the really unlikely hypothesis outweigh the more plausible one.

And, lo and behold, it turned out that Samuel Soal had been fiddling the numbers.

The trouble with the existence of multiple hypotheses is that it means that people with very different priors on something may never end up agreeing. If you think psychic powers are plausible, then your prior will of course be much greater than mine. If we both saw lots of evidence for psychic powers—for instance, someone guessing the right number between one and ten, a hundred times in a row—then, *assuming the only two possible hypotheses are fluke or telepathy*, our priors would be washed out by the flood of data. But if we have multiple hypotheses—for instance, fluke, telepathy, or fraud—then the evidence would render the fluke hypothesis irredeemably unlikely, but for you, it will make telepathy the most likely explanation, whereas for me it would put fraud in the top spot.

And that obviously has various real-world implications. Say you have some hypothesis, like "vaccines cause autism" or "man-made climate change is real," and you assign it some given prior probability. You take two people, one who thinks the hypothesis is likely, and the other who thinks it's not.

Then you give them some piece of evidence, like an article on the BBC News website saying, "These scientific studies show that autism rates didn't spike after the introduction of the MMR vaccine," or "these scientific studies show that the world is getting warmer and it roughly tracks atmospheric carbon dioxide concentrations."

Just as with the telepathy example, if the only possible explanation for the evidence is that the hypothesis is right (or wrong), then that evidence should help the two people's opinions converge. But there's an alternative hypothesis: that the source is untrustworthy. If one person strongly believes that MMR causes autism or that climate change isn't real, then the evidence will not bring them closer to the other person's beliefs, but will instead push them into saying, "This is why the BBC can't be trusted." (Or, if the BBC provides links to the scientific papers, "This is why the scientific establishment can't be trusted.") And, disturbingly, in many cases, that would be the rational thing to do.

BAYES IN AI

At its heart, artificial intelligence is just a program that tries to predict uncertain things. By this point, you won't be surprised when I say that it is fundamentally Bayesian. *Artificial Intelligence: A Modern Approach*, the standard textbook for undergraduate AI degrees, even has a picture of Thomas Bayes on the front cover, and says, "Bayes' rule underlies most modern approaches to uncertain reasoning in AI systems."[20]

There is such a thing as "Bayesian machine learning," which is *explicitly* Bayesian, its architecture designed intentionally to mimic Bayes' rule. I'm not talking about that. I'm saying that—because, as we've seen, Bayes' theorem essentially *is* decision-making—all machine learning/AI systems are Bayesian.[21]

Imagine a very simple AI that tries to identify pictures of rats, dogs, and lions. Really not that many years ago, it would have seemed astonishing, but nowadays it's not much to write home about. (In 2017, when I was doing the interviews for my first book, it was still pretty exciting that AIs could reliably tell a dog from a cat. Now you can ask your smartphone to search your camera roll for pictures of dogs, or of babies, or beaches, or whatever, and it will bring them all up in a fraction of a second.)

At a very abstracted level, here's what it does:

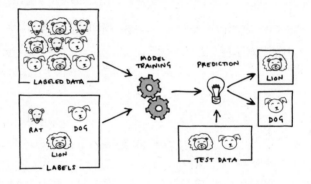

You give it however many thousands or millions of pictures of rats, dogs, and lions, each labeled as rat, dog, or lion, to train on (its "labeled data"). It sloshes them around in its circuits in some fashion, and then, once it's done that, you give it new pictures to identify (its "test data"). It will then label each of those pictures as rat, dog, or lion, according to its best guess. This model of AI is called "supervised learning." What it's doing is *predicting* what the humans who labeled the training data would label the test data.

Of course, there's a simple and almost tautological way in which this is Bayesian. Before seeing a picture, the AI presumably has a subjective prior probability of p ≈ 0.33 that it will be a lion: a one-in-three chance. After seeing it, and gaining the new information, it updates its probability to p = 0.99 or whatever. Prior, likelihood, posterior.

But we can be more specific than that. Let's simplify the situation even further and look at it as a graph. This even simpler AI is just looking at where a bunch of blobs are on a graph and trying to find the line of best fit through them. This isn't something you need a powerful AI for: it's just a linear regression, statistics that Francis Galton would have been entirely comfortable with. But it's the same principle.

Let's take a chart of people's shoe size versus their height. You take a large, random sample of people, measure their height and their feet, and plot them on a graph—foot size along the x-axis, height up the y-axis. As you'd expect, on average, taller people have larger feet, but there's some variation. So the dots tend to line up from bottom left to top right.

What your AI wants to do is draw a line through them. You could draw the line by eye, but there's an established system called the *line of least squares*. Draw a line on the graph and measure the vertical distance from the line to each dot. That distance is the *error*, or the *loss*. For each dot, square the error—that is, multiply it by itself, so that all the numbers are positive. (A negative number squared is positive.) Then add the squared error for all your dots together.

That figure is the *sum of squared error*. What you want to find is the line with the lowest squared error: the one that has the smallest average distance to all the dots.

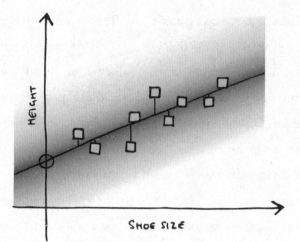

Those dots are your AI's training data. And, of course, it's a Bayesian process. It starts out with a flat line, a flat prior. Then as you add dots—your data—the line moves, to give you the posterior distribution, which becomes the prior for the next bit of data.

But now you want to use it to make predictions. Say you give it someone's shoe size, and you ask it to guess that person's height. All it needs to do is go along the x-axis to the relevant bit—size 11, say—and then go up on the graph to where the line of least squares is. That's its best guess for the person's height. How confident it is depends on how much training data it has and the variance in that data—if it's very spread out, then the guess will be less confident.

This is, roughly, what real AIs do. They do much more complicated versions, with hundreds or thousands of parameters instead of just "shoe size" and "height," but the basic idea is the same. They have some training data, and they use it to predict the value of some parameter or parameters, given some other parameter.

Often, an AI is trained just once, and the test data doesn't change its priors. But that doesn't need to be the case. It could easily be that the AI continues to update on each piece of training data. The shape of the line each time is its prior, the new data point is the likelihood, and together they make a new posterior probability. The farther each dot is from the line, compared to where it was predicted to be, the more the model is "surprised" by it and updates its prediction for the next one. (Imagine the line is fuzzy and gets fainter as you get farther from its center.)

So far, I've sort of assumed that the line is straight. But it doesn't need to be. Lots of charts would be best suited to curved lines. For instance, if you imagine the shape of a graph with "Cumulative number of COVID cases worldwide" on the y-axis, and "Time" along the x-axis, starting at November 2019, your line of best fit should be an exponential curve, as the number of cases doubled every few days. Other times, you might find that an S- or J-shaped curve best fits the data, or a sine wave, or all sorts of things. You could arbitrarily say that your model has to draw a straight line, but that will often be the wrong choice: it will be "underfitting" its curve to the data.

Equally, your AI could, assuming it is sophisticated enough,* simply draw a wibbly-wobbly line that goes perfectly through the heart of every dot in the training data. That would give it a squared error of precisely zero. But it probably wouldn't represent the real underlying cause of the data. When new data arrives, it'll most likely be a long way

---

* Specifically, in the case of a neural network, which is the basis of most modern AIs, it would need to have more "parameters," that is, links between nodes in the network, than there are data points. If that's the case, it can draw a line that wobbles through all of them.

away from the weird wibbly-wobbly line your AI has drawn, because it's "overfitted" to the data.

The question, therefore, is how much freedom the AI has to wobble the line around. That freedom is analogous to what in the last section we called "hyperparameters"—as well as the simple question of the best-fitting curve, there's a higher-level question of how wobbly that curve should be. The AI's prior beliefs about those parameters are its hyperpriors. And they often decide that, all else being equal, you choose the simpler of two lines. You trade off simplicity against fit. Just what we were saying in the section on Occam priors.

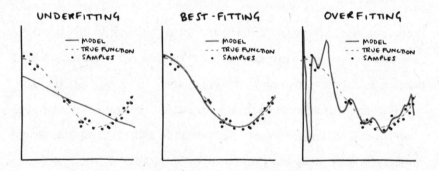

When a radiology AI tries to recognize cancers on a scan, or when ChatGPT writes a short story in the style of the King James Bible about a man getting a peanut butter sandwich trapped in his VCR, they're doing something Bayesian. They are using their training data to produce prior probabilities, which they then use to predict future data.

FROM AUTOCOMPLETE TO INTELLIGENCE

As we've just seen, AI is about prediction. Even the very fancy new AIs that have been making headlines lately are in a sense just "predicting" what a human would say or draw in response to a prompt.

Imagine ChatGPT, the "chatbot" released by the company OpenAI and used to power Microsoft's Bing search engine. It's trained, like the image-classifying AI in the previous section, on a vast dataset of text. That dataset is processed through the AI's neural network with its billions of parameters. And, in a broadly analogous way to that image classifier, it "predicts" what would tend to come after a text prompt. So if you ask ChatGPT, "How are you?" it might reply, "Very well thank you," not because it is, in fact, very well, but because the string of words "how are you" is often followed by the string "very well thank you."

What's been surprising about so-called large language models like ChatGPT is the extent to which that fairly basic-sounding ability to predict things actually leads to a very wide set of skills. "It can play very bad chess," says Murray Shanahan, an AI researcher at Imperial College London and DeepMind. "It doesn't do very well, but the fact that it can stick to the rules and play at all is surprising." What it's doing is predicting what follows strings of text like "e4," or "queen's knight to bishop 3." The prediction might be something like "e5," or "pawn takes knight."

As you can imagine, you can model this in a Bayesian way even if the AI isn't explicitly running Bayes under the hood. The prior probability of a randomly selected piece of text being "pawn takes knight" is fantastically low, but given the information that the previous piece of text was "queen's knight to bishop 3," the probability goes way up.

The fact that these AIs are predicting the next word (or "token," in the jargon) in a sequence means that a lot of people call them "fancy autocomplete" and say that it's not "real" intelligence. They don't "really" understand the world, they just mechanically predict that, statistically, "How are you" is often followed by "Fine thanks." Even if it looks much more complicated than that—say, following the prompt "Write me a short story set in the *Dune* universe in the style of P. G. Wodehouse" with a two-thousand-word piece about Duke Augustus "Gussy" Atreides trying to stop his aunt marrying him off to a Harkonnen—it is still, skeptics say, just predicting what comes next. A famous paper in 2021 called them "stochastic parrots" and claimed that language models work by "haphazardly stitching together sequences of linguistic forms . . . without any reference to meaning."[22]

But is this true? After all, one way to make good predictions is to build an accurate model of the world so that you can understand what is likely to happen in it. And, to some extent, that's what large language models (LLMs) appear to be doing. When you ask the LLM to write the story, does it have, in its neural network, some kind of model of the P. G. Wodehouse *Dune* universe, with its characters and its worlds, or is it just mechanically putting one word in front of another?

No one really knows, to be clear. LLMs, like all modern neural-network AIs, are far too big and complicated for humans to fully understand their workings. We can't see what's going on inside their "brains." But there is evidence.

"To what extent do LLMs emergently acquire models of the world through their objective of next-token prediction?" asks Shanahan. "Because having a world model might actually make it easier."

The human brain—as we'll see in a lot more detail later in the book—seems to do something like this. It builds a model of the world, uses that model to predict the signals coming in from the world, and updates its model according to how closely its predictions match reality. It's "just predicting" what the next signal will be, but in order to "just predict," it needs to *understand*. What we call "understanding" is, really, just having a model of something that successfully predicts its output.

Are LLMs doing the same thing? "The answer is almost certainly yes," says Shanahan. "But there are some nuances."

There was a study released in 2023 that used an AI that had a similar architecture to an LLM like GPT, but was instead trained on the board game Othello, a Go-like game in which players take turns to place small black or white disks on an eight-by-eight board.[23] All it saw was game notation from tens of thousands of games, tokens noting where each player placed their disks. The idea was to see whether the AI simply memorized a load of statistics about the game—noticing that "f5" is often followed by "d6" in the notation, for instance—or whether it built an internal representation of the board. That would be a "world model," albeit only of the small, limited world to which the AI had access.

They then looked to see if the AI would make original, legal moves that it had never seen before—by artificially restricting its training data, removing all games starting with one of the four possible moves, so that they knew that all games starting with that move would be novel to the AI. And yet it still made very few errors—only one of its moves in every five thousand was illegal. It was clearly not simply memorizing the statistical correlations.

Then they used a technique called "probing" to look at the internal state of the AI at certain points, and see if they could use it to predict the board state of the game at that moment. They could, with considerable accuracy—which showed that the AI had created some sort of representation of the board. Then they changed those internal states manually, so that its model of the board would be different, and the AI made moves that would only be legal in those board states, implying that it was *using* that internal representation to make decisions.

It's as though, writes Kenneth Li, one of the authors of the study, you played Othello with a friend every day.[24] "The two of you take the competition seriously and are silent during the game except to call out each move as you make it, using standard Othello notation," he says. But as you do so, a crow sits outside the window, out of sight of the board. And after a while, "the crow starts calling out moves of its own—and to your surprise, those moves are almost always legal given the current board."

You look outside and see that the crow has scratched an Othello board into the dirt and is using seeds to represent the disks in your current game, that it at least has a model of the board. And then you move one of the disks around, putting the game in a different position, and the crow makes a new, legal move from that position. "It seems fair to conclude the crow is relying on more than surface statistics," says Li. "It evidently has formed a model of the game it has been hearing about."

Othello-GPT uses the same basic architecture as ChatGPT and the other LLMs, so it's reasonable to assume that these findings tell us something about all of them. But in case you're not convinced, another paper, from 2021, looked at language models directly, and found that if you take narrative descriptions like "You see an open chest, containing nothing but an old key," and tell the AI you've taken the key out, it works out things like "the chest is empty." "Probing" of its internal state, like that done with the Othello-GPT, showed that once again it had an internal representation and used it to make decisions.[25]

As we'll see in the last section of this book, in quite a real sense, all humans are doing is "predicting" the world around us. But in order to do so, we build a sophisticated, rich model of that world, to help us make *good* predictions. The suggestion is not that ChatGPT, or Othello-GPT, understands the world in anything like the way that we do. But it does show that *just predicting* is more than just autocomplete. Prediction—which, remember, is an inherently Bayesian process—gets you a long way toward intelligence.

CHAPTER FOUR

Bayes in the World

ARE HUMANS IRRATIONAL?

In the last chapter, we talked about how Bayes' theorem is the ideal form of decision-making. If you could take account of all the information available to you, you could assign prior probabilities in an optimal way, and update them appropriately as new information came in. Actually doing so is impossible. But we must be doing *something* right, in order to be making decisions at all. But how good are we as Bayesians?

Over the last few decades, there's been a lot of research into how irrational we are.* The most famous work is probably that of Daniel Kahneman and Amos Tversky, a pair of Israeli psychologists—Kahneman

---

*When decision theorists call something "rational," they mean that it is the most likely way to achieve some goal. It doesn't matter whether your goal is making money, or achieving world peace, or building a ninety-foot-tall tower of used chewing gum. You can behave "rationally" in this sense even while doing things most of us would call very stupid.

would win the Nobel Prize for economics for it (Tversky had died by that point, and Nobels are never given posthumously).

For instance, research has shown that in certain situations we are not very good at judging risks. According to a famous 1978 study,[1] if someone asks us to guess how likely we are to suffer some bad thing, instead of trying to think of base rates and population prevalence, we tend to answer an easier question instead, such as "How easily can I think of an example?" That's called the *availability heuristic*, and it's why we tend to think of dramatic, memorable, newsworthy risks as more common than boring ones—people think terrorism kills more people than domestic accidents, or Ebola is more dangerous than diabetes, and they're off by several orders of magnitude.

We make basic logical mistakes like thinking that "Björn Borg will lose the first set" is less likely than "Björn Borg will lose the first set but win the match," even though it's logically impossible for Björn Borg to lose the first set but win the match without losing the first set first. Similarly, people rate "Reagan will provide federal support for unwed mothers and cut federal support to local governments" as more likely than "Reagan will provide federal support for unwed mothers," even though Reagan can't do both things without doing the first thing. (These examples are taken from a 1981 study by Kahneman and Tversky.)[2]

That's because, in decision theory, the stuff we were talking about in the last chapter, the probability of both A and B happening must by logical necessity be smaller than, or at most equal to, the probability of either A or B happening on its own—in notation, $(P(A,B) \leq P(A))$. The number of universes in which Björn Borg loses the first set cannot be smaller than the number of universes in which Björn Borg loses the first set but goes on to win the match. Misunderstanding that is called the *conjunction fallacy*.

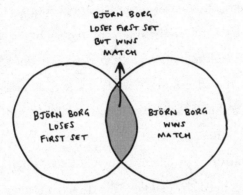

People also get confused by *framing effects*, Tversky and Kahneman found in another 1981 paper.[3] Say you tell people that there has been an outbreak of a novel disease that is expected to kill six hundred people and that there are two possible approaches to combating the disease, one reliable but partial, the other a long shot but potentially perfect. Tversky and Kahneman found that if people were told that the first program would definitely save two hundred people, and that the second program would have a one-third chance of saving all six hundred, but a two-thirds chance of saving none, then nearly three-quarters of respondents chose the sure thing. But if you reversed the framing—told people that the first program would definitely mean four hundred people would die, while the second program meant a one-third chance that nobody would die, and a two-thirds chance that all six hundred would—then the respondent numbers reversed too. More than 75 percent of people chose the gamble.

Of course, these two framings are logically equivalent—"four hundred will die" is the same as "two hundred will be saved." But the way they were presented entirely changed people's approaches to them.

Findings like these laid the foundation for a miniature publishing industry of "aren't humans all mega-irrational" books, notably *Predictably Irrational* by Dan Ariely and *Irrationality* by Stuart Sutherland, and to an extent Kahneman's own *Thinking, Fast and Slow.*

It's not that the theses of these books were wrong—most of this research does stand up to scrutiny, even after 2011 and Daryl Bem and all those things, knowing what we know now about the replication crisis and the statistical problems in psychology. When presented with questions framed like this, people really do seem to give incoherent, irrational answers. Dan Ariely's own work has come under scrutiny after a 2012 paper of his[4] turned out to be based on fraudulent data—Ariely denies making up the data himself, but admits he has no good story for how it happened.[5] And a lot of the work on "social priming" that Kahneman's book cited has since been undermined, as we discussed in the section on the replication crisis in chapter 2. But it's definitely true that framing affects how people view risk, and that people misjudge risk on the basis of how easily they can think of examples.

People are also bad at explicitly working out how to incorporate prior probabilities and new evidence—at being conscious Bayesians, in other words. That's true even of people who really ought to be doing better at it. A famous 1978 study asked sixty medics—twenty medical students, twenty junior doctors, and twenty more senior doctors—at Harvard Medical School the following question: "If a test to detect a disease whose prevalence is 1/1000 has a false-positive rate of 5%, what is the chance that a person found to have a positive result actually has the disease, assuming you know nothing about the person's symptoms or signs?"[6]

As you'll know from having read this far, it's pretty easy to work out. I tend to do it by imagining a much larger group—say a million. Of the million, 1,000 will have the disease and 999,000 won't. Of the 999,000, our test will return false positives on 49,950. So assuming it correctly identifies all 1,000 who do, anyone who has a positive test will have a slightly less than 2 percent chance of having the disease (1,000/(49,950 + 1000) ≈ 0.02).

This seems like something important for doctors to be able to work out. But the 1978 study found that only eleven of the sixty medics gave the right answer (and those eleven were evenly spread among the groups: the students did no worse than the senior doctors). Nearly half said 95 percent: that is, they failed to take base rates into account at all.

Other studies have found similar results. A 2011 paper asked junior obstetrics and gynecology doctors: "Ten out of every 1,000 women have breast cancer. Of these 10 women with breast cancer, 9 test positive. Of the 990 women without cancer, about 89 nevertheless test positive. A woman tests positive and wants to know whether she has breast cancer for sure, or at least what the chances are. What is the best answer?"[7]

That version of the question really walks you through it—there are nine true positives and eighty-nine false positives! All you have to do is work out what nine is as a proportion of nine plus eighty-nine. Nonetheless, of the nearly five thousand residents who answered the question, just 26 percent got it right.

I used to think the takeaway from all this was that humans are deeply irrational. I've moved somewhat away from that, though. Since we know that ideal decision-making is necessarily Bayesian, and since humans mostly make good decisions—most of the time, we successfully find food to eat, seek shelter to get out of the rain, and avoid being hit by cars—we must be doing something right. And I think a lot of the "Humans are so biased!" discourse is usually really saying, "*Other* humans are so biased."

The way to think about it is that we humans are amazingly rational, *if information is presented to us in ways that we are designed to process it.*

That's Jens Koed Madsen's position, anyway. He studies human rationality in his work as a psychologist at the London School of Economics. "If you've sat there and fiendishly designed a behavior experiment, and it's taken you two months to design," he says, "maybe it's not all that much like our day-to-day lives. Maybe it's artificial. If you look at people in their everyday lives, they're kind of fine, in like 90 percent of their decisions. If I want to buy a coffee, I'm capable of going to get one from the café."

He uses a different example. There's another famous experiment designed to show how silly we all are, called Wason's selection task, devised by Peter Wason in 1966.[8] Here's a version of it:

There are four cards on the table. Each one has either a number or a letter on one side, and either a person or an animal on the other side. The four faces you can see are a star, an 8, a young woman, and a rabbit. You're told that if a card shows a number on one side, the other side shows an animal. Which cards do you have to turn over to find out whether that's true?

Have a think for a second!

I'm just going to leave some line breaks so that hopefully you can think about it without accidentally seeing the answer.

. . . tum te tum . . .

. . . OK. So. Most people turn over the 8 and the rabbit. Which makes intuitive sense, since the claim is about squares and animals, but it's wrong. The correct answer is the 8 and the young woman.

This is pure Aristotle-and-George-Boole-style propositional logic. The claim is "*If* X (one side of the card has a number) is true, *then* Y (the other side has an animal) is true."

You can show that claim is false in two ways. One, you could look at an example where you know X is true and find that Y is not true; or you could look at an example where you know Y is not true and find that X is true. So you look at X and at not-Y.

You can turn over the 8, and if there's no animal on the other side, you know that the claim "If number, then animal" is false. (X but not-Y.)

But also, if you turn over a non-animal picture, and find that it has a number on the other side, then that also proves that the claim "If number, then animal" is false. (Not-Y and yet X.)

Meanwhile, if you turn over a picture of an animal, and it doesn't have a number, that doesn't prove anything. "If X, then Y" isn't disproved by "Y and yet not-X."

If you got it wrong, don't worry. I have seen that question about a dozen times and always struggle to piece it together, even knowing that it's a trick and that all I have to do is remember exactly what the trick is. In Wason's original study he found that less than 10 percent of respondents got it right, and later replications found similar results.[9]

The usual write-up of this is that it shows that humans are highly prone to confirmation bias—that instead of looking for ways of falsifying a hypothesis, we look for evidence to support it, so that our preexisting beliefs aren't challenged.

But Madsen thinks this is something of a gotcha. "Think back to your student days," he says. "You're at a college party, and you know it's illegal for the party to be serving alcohol to minors." (I should say that I was twenty when I went to university, and my excessive drinking was perfectly legal by UK law, but I'm going to pretend I'm being asked the same question about high school.)

Madsen imagines the following situation: The campus police are coming. You see four friends of yours, all drinking. You know the ages of two of them—one's twenty-one and one's sixteen—but you don't know what they're drinking. And you can see what the other two are drinking—one's got an orange juice and one's got a beer—but you don't know how old they are.

As you can probably see, that's the exact analogue of the Wason selection task. You could imagine it with four cards, like this:

Obviously, you need to check the age of the kid who's drinking beer, and the drink of the kid who's sixteen. "In that situation, everyone gets it right," says Madsen. "If the twenty-one-year-old is hammering tequila, fine. If the sixteen-year-old is drinking orange juice, fine." This isn't just Madsen's hunch. He's used these examples on his students and finds they all get it, when it's put in concrete terms like this: "They'll just check those two people to make sure the party isn't shut down." A 1992 study along these lines by two evolutionary psychologists also found that 75 percent of respondents got it right,[10] compared to less than a quarter given the same logical problem with more abstract terms.

"The selection task is mentioned in every textbook to show how irrational we are," says Madsen. "But are we really, if we place it in a natural ecological environment?

"If you require that we phrase it in a super-abstract way to get an effect, is there really an effect, or is it only in the margin? It seems unkind to people to say, 'Because you couldn't do this very abstract task that I've labeled specially to be super-tricky, then you are full of confirmation bias and failed to falsify things.' Especially when people can then come along and do it a hundred times out of a hundred in a natural environment."

This seems to be true: humans are genuinely pretty good at reasoning, when the reasoning takes place in a format familiar to them. Steven Pinker recounts a story of a southern African hunter-gatherer tribe, the San, also known as Bushmen, which he takes from the research of an anthropologist called Louis Liebenberg.[11] You might not instinctively think of southern African hunter-gatherers as practitioners of Bayesian reasoning, but Pinker argues that they are.

The heel of a porcupine's paw has two pads (the "proximal" pads, near to the arm, as opposed to the intermediate pads, where the palm of your hand would be, or the digital pads, on the claws). A honey badger's paw has one proximal pad. Usually, a paw print will show all the pads, but sometimes—on hard ground, say—the print will be only partially visible. The San distinguish between the chance that a honey badger will leave a one-pad print—the sampling probability—and the chance that a one-pad print has been left by a honey badger—the inverse probability. A one-pad print might be an imperfect porcupine print. Further, the San will take into account the prior probability: if they find an ambiguous print, they'll think it's more likely to come from a common animal than a rare one. This is precisely how Bayesian reasoning works.

In modern life, we're often pretty good too. Another common "Aren't humans irrational?" gotcha is that you can give people the same speech, and tell them it was given by a politician they like or a politician they don't like, and their reaction to it will be entirely different.

But, points out Madsen, that's thinking about things like a frequentist—assuming that we can only use the current evidence to make decisions with. In fact, it's perfectly rational to take your prior beliefs about a politician's probity into account when assessing their policies. He and colleagues published a paper in 2016 asking American voters whether they thought a policy was likely to be good if a particular politician supported or attacked it.[12] The politicians they named were the five most high-profile candidates in the presidential primaries—Hillary Clinton and Bernie Sanders for the Democrats; Jeb Bush, Marco Rubio, and Donald Trump for the Republicans.

The authors got 252 people to rate the candidates on trustworthiness and political expertise. (In case you're interested, Sanders got the highest average for trustworthiness, Clinton the highest for expertise, and Trump the lowest for both.) Then they asked them about their opinion on a hypothetical, unspecified policy that politician either supported or attacked. Unsurprisingly, they found that people's prior beliefs about a politician's trustworthiness influenced whether they thought a policy that politician supported (or attacked) would be any good. But more interestingly, they found that people responded in a highly Bayesian way—the degree to which their prior beliefs influenced their posterior ones very closely matched an explicitly Bayesian model.

"It means that people are saying, 'I don't trust him, I think he'll have a bad policy,'" says Madsen. "And that's not irrational! It's just saying you have different views on people."

In general, it's best to think of human biases as products of our mental heuristics, shortcuts that allow us to do what would otherwise be extraordinarily complex math. The availability heuristic, mentioned above, probably works pretty well most of the time—it might fall apart when we're thinking about high-profile risks like school shootings, but if we're thinking about stuff in our daily life, like "How likely am I to get in trouble if I go through this red light on my bike?" it's a lot easier and more efficient than (and probably nearly as accurate as) trying to assess base rates and so on. A straightforward example is catching a thrown ball. Doing the math—assessing the likely path of the ball's parabola, moving to the place it will hit the ground, and positioning your hand and closing it in the exact timing—would be astonishingly complicated. To quote Douglas Adams:

A ball flying through the air is responding to the force and direction with which it was thrown, the action of gravity, the friction of the air which it must expend its energy on overcoming, the turbulence of the air around its surface, and the rate and direction of the ball's spin. And yet, someone who might have difficulty consciously trying to work out what $3 \times 4 \times 5$ comes to would have no trouble in doing differential calculus and a whole host of related calculations so astoundingly fast that they can actually catch a flying ball.[13]

But that's not what we're doing. When a cricket fielder sees the ball going high toward the boundary rope and starts running to catch it, they're not doing differential calculus. What they're doing is using a simple shortcut called the *gaze heuristic*. The psychologist Gerd Gigerenzer describes it like this: "Fix your gaze on the ball, start running, and adjust your running speed so that the angle of gaze remains constant."[14] No mathematics are involved at all. Experiments show that non-human animals use the same system—dogs catch Frisbees by keeping them in the same place in their eyeline,[15] just as baseball outfielders do to catch fly balls.

Experimenters can tell this is what catchers are doing, because if they had immediately solved the ballistic equations to predict where the ball would land, they would have run at full speed in a straight line to where the ball would end up, and wait there. But instead, they change running speed as they travel to keep the ball at the same angle in their eyeline, and slightly curve their run as they go.

The gaze heuristic is almost as accurate as actually calculating the trajectory, but computationally far simpler. The Royal Air Force realized during the Second World War that they could use the gaze heuristic to help guide fighters to intercept bombers, and that it would be much quicker than doing the math. Guided missiles such as the AIM-9 Sidewinder use it to shoot down enemy planes.[16]

When humans make decisions under uncertainty, they're doing something similar: using simple heuristics that are much less effortful and time-consuming than doing the conditional-probability mathematics. Sometimes, and especially under artificial laboratory conditions, those heuristics misfire and mislead us, and then we call them "biases."

"There are tons of them," says Madsen. "If you observe something new, you say you've found a new heuristic. But there's no overarching theory. It's like biology before Darwin. And they contradict each other." For instance: there's *recency bias*, which is that we overweight more recent evidence. But there's also *anchoring*, which is that the first thing we see tends to set our expectations. And there's *frequency bias*, the thing you've seen most often. "Recency is what you saw recently, anchoring is what you saw first, and frequency is what you saw most, and [yet] your decision theory uses all of them?

"It's not that we're always rational—maybe there are ditches that we fall in," he says. "Maybe confirmation bias is a really deep one, or conjunction bias is more shallow. But it's a question of how much these things influence behavior at the margins. Maybe we're super-irrational, walking around being wrong all the time, but I don't think that's true."

There obviously are specific ways in which humans make irrational choices: the classic example is that in the months following the September 11 terrorist attacks, Americans chose to drive long-distance journeys much more, out of fear of flying. A 2009 paper[17] found that that may have led to about 2,300 extra driving deaths—about two-thirds the death toll of the atrocity itself—because flying is so much safer than driving. Perhaps this is just my own political bias speaking, but I'd also say that the decision to spend trillions on invading Iraq and Afghanistan to reduce an already tiny but highly visible risk of terrorism, and spend next to nothing reducing the much greater but harder-to-picture risk of global pandemics, was irrational at the time and has been shown to be in hindsight.

But making decisions under uncertainty is hard. We don't have access to all the information, and integrating all the information we do have into the Bayes equation would be computationally impossible. Instead we use shortcuts and heuristics. But it seems that our instinctive decision-making, from a Bayesian perspective, isn't that bad.

Even with things like, say, vaccine hesitancy—if you have low trust in public health systems, then your prior will be to distrust vaccines, and you won't update much on new evidence coming from public health experts. That is, given your priors, perfectly rational. If someone wanted to persuade you otherwise, they would do better to build your trust in public health systems, rather than to provide you with a list of public health experts saying that vaccines are safe. "Maybe your assignment of priors is fucked up," says Madsen. "Fine. But that's where Bayes comes in as a lovely tool, because we have to understand that.

"What you find in Bayesian studies," he says, "is that, sadly or happily, people are a bit off, but basically reasonable. They're kind of fine. And that ain't very sexy, it won't sell a lot of airport books, but I think it's kind of sexy in its unsexiness. It's a way of saying that we're reasonable people. If we were completely all over the place, like the heuristics and biases people sometimes suggest, then how on earth would we have done what we've done? Build complex systems and skyscrapers and so on? By and large, we have gray areas and weak spots, but we're kind of OK."

It is, though, nonetheless true that when it comes to *explicit* assessments of probability, humans aren't always that good. In the next section, we'll look at that, and later, at some people who are trying to be better.

MONTY GOT A RAW DEAL

Humans can be good at implicit, automatic reasoning, as we have seen. But it's definitely true that when the situation calls for more formal, explicit probabilistic thinking, our heuristics lead us astray, to the point that some people—even people who definitely should know better—simply cannot accept the real answer, even when it's demonstrated to them very clearly.

There's a game show on US television called *Let's Make a Deal*. Contestants have to make a series of deals with the host. In its original version, the host was a guy called Monty Hall. The deals were like "Would you rather take [known sum of money] or [unknown gift in a box]?" All very Bayesian-decision-making-under-uncertainty.

Let's Make a Deal inspired a letter to the magazine *American Statistician*, in 1975.[18] The statistician Steve Selvin of the University of California at Berkeley imagined the following scenario. Monty offers a contestant three boxes: A, B, and C. In one of the boxes are the keys to a Lincoln Continental, which I assume is some kind of large car, and if you choose that box you get to keep the car as well as the keys. The other two boxes are empty.

The contestant chooses box B. Monty offers the contestant $100 for the box. Those of you who've been keeping up will think, hang on, a one-in-three chance of a car worth about $10,000 has a higher expected value than a sure thing of $100, and the contestant agrees: he turns down Monty's offer. Monty ups his offer to $500, and the contestant stays firm.

But then it gets interesting. Monty opens one of the *remaining* boxes on the table, box A. It's empty. And then he says: "Now either box C or your box B contains the car keys. Since there are two boxes left, the probability of your box containing the keys is now ½. I'll give you $1,000 cash for your box."

The contestant turns down that deal, but—to Monty's surprise—makes another one: "I'll trade you my box B for the box C on the table." Because, the contestant realizes, the probability of box B containing the keys isn't ½: it's ⅓. If he switches boxes, he'll have a two-thirds chance of winning.

The puzzle became famous in 1990, when it was published as a letter to *Parade* magazine columnist Marilyn vos Savant (apparently the "smartest woman in the world," with a recorded IQ of 230). By then, the puzzle had changed its format a little, but the underlying structure was the same. Here's how it was presented to vos Savant: "Suppose you're on a game show, and you're given the choice of three doors. Behind one door is a car, behind the others, goats. You pick a door, say #1, and the host, who knows what's behind the doors, opens another door, say #3, which has a goat. He says to you, 'Do you want to pick door #2?' Is it to your advantage to switch your choice of doors?"[19]

Vos Savant had no doubt about it. She agreed with Selvin: the right thing to do is to switch. If you keep the door you originally chose, there's a ⅓ chance you'll get the car, but if you switch it's ⅔.

Does that seem weird to you? It does to most people. It made an awful lot of vos Savant's readers very angry, including several mathematics PhDs, such as this guy: "You blew it! . . . As a professional mathematician, I'm very concerned with the general public's lack of mathematical skills. Please help by confessing your error and in the future being more careful." And this one: "May I suggest that you obtain and refer to a standard textbook on probability before you try to answer a question of this type again?"

It also confused Paul Erdös, one of the greatest mathematicians of the twentieth century, who insisted when presented with the problem: "That is impossible. It should make no difference if you switch."[20] Most people think the true probability is 50 percent, 0.5, and it makes no difference if you switch or not.

But most people are wrong. The angry mathematics PhDs were wrong, Erdös was wrong, and vos Savant and Selvin were right. Assuming that Monty knows which door the car is behind, and that he always opens one of the other ones, you should switch.

There are a few ways to grasp this intuitively. One is to imagine that, instead of a choice of three doors, you'd originally had a choice of 1 million doors. Once again, one of them hides a car. (And you would like a car.) You choose one. Then Monty opens 999,998 doors, showing that all of them are empty. Now you're left with just two, your original choice and one other.

You can also think of it like this. Before Monty opens the door, there's a one-in-three chance that you picked the correct one. In that case, it would be bad if you switched. But there's a two-in-three chance that you picked a wrong one, and in that case, it would be good if you switched.

Or imagine that you play the game three hundred times. In one hundred of those, you pick the right door, and switching means that you lose. But in two hundred, you pick the wrong door, and switching means that you win.

But what's crucial is that Monty, first, *knows where the car is*, and second, *always opens a door with a goat behind it*. Once that's stipulated—or assumed—you can easily work out the odds with Bayes' rule.

At the start, your probability that the car is behind any given door is one-third, or $p \approx 0.33$. You have no information that gives you any reason to pick any one over the others. But Monty, if he knows where the car is and always opens a door, gives you some information; he allows you to update your prior probabilities.

It's easier if we use odds. The odds of the car being behind Door #1, #2, or #3 are 1:1:1: that is, there's nothing between them. That's still true even after you pick Door #1.

Then Monty opens Door #3 and reveals a goat. The probability that he'd open Door #3 if you're right and the car is behind Door #1 is 50 percent—he could have picked either of the other doors. But if the car is behind Door #2, then it's 100 percent certain that he'd have picked Door #3. And if it was behind Door #3, there's a 0 percent chance. So the odds are 1:2:0. That's your likelihood, or your Bayes factor.

You might just remember that when you're doing Bayes' theorem with odds, it's nice and easy: you just multiply your prior odds with your Bayes factor and you get your posterior odds. 1:1:1 times 1:2:0 is 1:2:0. You know it's not in #3, and it's twice as likely to be in #2 as #1.

But now imagine a different scenario, where Monty doesn't always open a wrong door, or he doesn't always open a door at all. Or if you don't know if Monty has any particular strategy. Imagine that Monty actually flips a coin, and if it comes up heads, he opens the lower-numbered of the two remaining doors. Or Monty's not even doing it: an earthquake hits the TV studio just after you make your decision, and one of the two remaining doors happens to swing open.

Then you're not gaining any information about the door you picked. The prior odds are 1:1:1, but the chance that Door #2 would be the one that happened to open is fifty-fifty whether or not you picked the right one, so your likelihood odds are 1:1:0 and therefore your posterior odds are 1:1 as well. Then it really doesn't make any difference whether you stick or twist. Of course, there was a 50 percent chance that the earthquake or Monty's coin toss would have revealed the car— you were unlucky there—but it didn't.

Again, imagine it being played three hundred times. In one hundred, you pick the right one, then Monty flips the coin, and opens one of the two remaining empty doors. You switch and lose out. In the other two hundred, you pick the wrong one. In one hundred of those, Monty opens the other wrong door. If you switch, you win. But in the remaining one hundred, Monty opens the door with the car behind it, and the game is over. So in the two hundred universes where the game is still going on, there are one hundred in which switching is good, and one hundred in which switching is bad.

Bayesian reasoning requires you to use *all* the information at your disposal. You don't just know that Monty opened a door: you know (or you have some reason to believe) his *algorithm* for opening that particular door. You know why he did it. And that information changes your beliefs, and therefore your estimate of the probability that the car is behind any particular door. But that feels weird to us. Just as it feels weird that you can get a positive result on a 95 percent accurate test and still only have a 2 percent chance of having the disease it tests for.

This all gets much weirder (or it does for me) in the second example, the "Boy-Girl paradox," created by the American science popularizer Martin Gardner in 1959.[21] Imagine you meet a mathematician, and they tell you that they have two children. You ask if at least one of them is a boy. (It's a strange question, but this problem is incredibly sensitive to tiny changes in wording, so I have to be careful.) The mathematician says, yes, at least one of their children is a boy. What's the chance that they are the parent of two boys?

It *obviously should be fifty-fifty*. The other one is a girl or a boy! It doesn't matter what the one you know about is! But . . . it's not. The chance is one-third again.

As you may be able to tell, this drives me crazy. But it's very much unavoidable. Just as Fermat and Pascal realized nearly four hundred years ago, what matters is the number of possible outcomes (assuming that all those outcomes are equally likely). There are four possible pairs of children that a parent of two might have: girl-girl, girl-boy, boy-girl, and boy-boy. They're all, to a first approximation, equally probable.

If you know that *at least one* of the children is a boy, but you don't know which one, then you've ruled out one of those combinations—girl-girl. Girl-boy, boy-girl, and boy-boy all remain. You've already got one boy in the bank, as it were. So the other child is either a girl or a boy. The unknown child is twice as likely to be a girl as to be a boy. (Which I find weird because it's like there's some strange quantum effect where knowledge of one child affects the sex of the other.)

Once again, this is sensitive to far more than just the bare fact that there's a boy. If you know that the *elder* child is a boy, you've ruled out two possibilities—girl-girl and girl-boy. So now there are only two remaining outcomes, boy-girl and boy-boy. The posterior probability of the other child being a boy is 50 percent.

Or imagine if the mathematician came up to you and said, unprompted, that they have two children and one of them is a boy. You would probably not think, "And it's completely unknowable to me what the other one is!" I at least would assume that the other is a girl (why not say, "They're both boys?").

The math of this is entirely the same as the Monty Hall problem above. But for some reason I find it far more counterintuitive, and I wanted to share the frustration with you all. And I'm not alone, I think. People struggle with this sort of thing: explicit probability-reasoning, actually dealing in numbers and percentages rather than instinctive catching-the-ball-style heuristics. Next, I want to talk about some people who try to be better at it.

SUPERFORECASTING, PART 1

I love this story, and I tell it all the time, so if you've ever read anything by me before, you may have already heard a version of it. Apologies in advance.

In 1984, the Cold War was feeling pretty hot. The USSR and the US had stockpiled vast numbers of nuclear weapons; tensions were high. I was born in 1980, and for much of my childhood there was a sort of background assumption that nuclear war was coming. *Threads* on the BBC, *When the Wind Blows*, Sting asking whether the Russians love their children too; the art of the day was written, to quote Queen, in the shadow of the mushroom cloud.

The tension was partly caused by uncertainty. The Soviet premier, Leonid Brezhnev, who had been in power since 1964, had died in office in 1982. He had been replaced with Yuri Andropov, who was already sixty-eight at the time and somewhat unwell. He suffered a kidney failure in early 1983, and died a year later, having spent much of the intervening time semiconscious in hospital. He was replaced by Konstantin Chernenko—"an enfeebled geriatric so zombie-like as to be beyond assessing intelligence reports," according to one historian[22]—who was widely expected to die soon too. Ronald Reagan's presidential administration was struggling to build relationships, and it was widely expected that another Soviet hard-liner would follow Chernenko.

The tension was so high that awful mistakes were made. In September, a South Korean airliner, Korean Air Lines flight 007, strayed into Soviet airspace. Moscow dispatched fighters to intercept it and mistakenly shot it down; among the 268 dead was an American congressman.

And then the world came awfully close to nuclear war. Each autumn, NATO would carry out a war game called Able Archer, preparing Western forces for the buildup to and the event of an all-out nuclear attack. The 1983 version was in November. But it was more comprehensive than most years, and involved more realistic communications and the participation of heads of government. Moscow saw it taking place, and convinced itself that it was part of a ruse to cover a genuine attack. They started loading nuclear warheads onto bombers. Only a "well-placed spy" in the KGB's London headquarters, who passed the info on to Washington via British intelligence, alerted the White House to how close they had come to accidentally sparking a holocaust.[23]

In this tense atmosphere, the National Research Council, part of the US National Academies of Sciences, received a grant to put together a panel tasked with the job of "preventing nuclear war." The panel included some very high-profile researchers. Amos Tversky, whom we know for his research with Daniel Kahneman, was among them. Three other panel members had already, by that point, won a Nobel. Others were high-ranking military officers, Kremlinologists, government officials. But one was, in his own words, "by far the least impressive member of the panel"[24]—a thirty-year-old freshly minted associate professor of political psychology at UC Berkeley called Philip Tetlock.

While Tetlock was there, he noticed something. Everyone agreed that after his inevitable and imminent death Chernenko would be replaced by another grim-faced politburo lifer—but they disagreed as to why. The liberal-minded in the group thought that the tough anti-Soviet policies Reagan favored were strengthening the hard-liners in the Kremlin and making reform impossible. The conservatives thought that the Soviet system was designed to produce repressive authoritarians, and so it would produce another repressive authoritarian. "They were equally confident in their views," Tetlock wrote.

They were right that Chernenko would die soon: he lasted barely a year after taking office. But then something happened that neither liberals nor conservatives expected. The politburo appointed Mikhail Gorbachev, youngish at fifty-four, energetic, and a committed reformer. Immediately, Gorbachev got to work—he imposed new policies of *glasnost*, meaning "openness," and *perestroika*, meaning "restructuring." He made a concerted effort to reach out to Reagan and the US, which Reagan cautiously but gladly accepted. Within months, the two leaders were talking about disarmament.

Neither the liberals nor the conservatives had predicted this. But Tetlock noticed something: both groups seemed to think it made perfect sense given what they already believed, and that they knew what was going to happen next, despite not having predicted this at all. The liberals thought that Reagan deserved no credit at all: it was down to a new era of Soviet leaders, tired of watching the USSR's economy crumble. Conservatives thought it was down to Reagan forcing the arms race to the point where the Soviets could no longer keep up and had to back out of the competition. Both sides, in short, thought that the completely unforeseen events proved that they had been right all along. Tetlock thought that maybe *whatever* had happened, people would have done the same thing.

A few years later, Tetlock set up a new study. He wanted to test the judgment of all these experts. Not that he doubted their intelligence or their integrity, but he thought that perhaps everyone, when confronted with unexpected information, found ways of saying that it just showed that whatever they already believed must be true.

Tetlock gathered 284 experts—journalists, military leaders, politicians, academics—and asked them to make more than thirty thousand predictions. Crucially, the predictions were falsifiable and time-limited: they were asked things like "Will the yen be higher than it is now against the deutschmark in one month's time?" And they put numbers on those predictions.

This was to avoid what Tetlock called "vague verbiage." If someone says something is *likely*, or that it *may happen*, it's not clear what they actually mean by that. Do *likely* things happen 30 percent of the time or 60 percent? If something *might happen*, does that mean it's 5 percent likely or 50? Research has found that people use these words in very different ways:[25] the phrase "real possibility" can mean 20 percent or 80 percent depending on who says it. And, of course, you're not really tied to any outcome. If it happens, you can say you predicted it, and if it doesn't, you can say you only ever said it *could* happen.

"I can confidently forecast that the Earth may be attacked by aliens tomorrow," Tetlock writes. "And if it isn't? I'm not wrong. Every 'may' is accompanied by an asterisk and the words 'or may not' are buried in the fine print."[26]

Instead, he asked everyone to give precise numbers. It's 45 percent likely that Greece's sovereign wealth fund will default this year, or there's a 10 percent chance that fighting between North Korea and South Korea kills more than one hundred people before 2030. And then—over the following months and years—they saw how many of those predictions came true.

Here's the clever bit, though. Each forecaster was asked to make around one hundred forecasts each. Some of them would have been 80 percent, some of them 40 percent, and so on. ("Walter Mondale to win the Democratic primary, 65 percent," or whatever.)

At the end of the study, they looked at how many of the forecasts came true. If 60 percent of your 60-percent-likely forecasts came true, and 30 percent of your 30 percent forecasts, etc., then you were *well calibrated*. Your assessments of the probability of things were good. If they came true more often than you'd thought, then you were *under-confident*. If they came true less often, you were *overconfident*.

Of course, there's more to good forecasting than calibration. If you predicted "50 percent likely" every single time, then, depending on the questions, you might do extremely well on calibration. But you'd be precisely no use as a forecaster—you'd provide zero information.

So Tetlock's study also rewarded (justified) precision. If you made a 90 percent forecast and it came true, then you got more points than for a 60 percent forecast. But if you made a 90 percent forecast and it *didn't* come true, you lost more points.

THE BRIER SCORE

Tetlock used something called the *Brier score* to assess his pundits' abilities. Brier scores were developed in weather forecasting, to assess the accuracy of previous forecasts. The lower your Brier score is, the better your forecast.

How they work is that they take the *squared error* of your forecast.

After a forecast comes true (or doesn't), that forecast has a probability of one (or zero). The error is the difference between that and your prediction. So if you said it was 80 percent likely that you would be on time for work, you'd write that as 0.8. If you then were on time for work, you'd subtract that 0.8 from 1, and you'd have an error of 0.2. Then you square that, and get 0.04. If you weren't on time for work, you'd subtract it from 0, and you'd have an error of −0.8. Whatever your error is, you square it. So your −0.8 becomes 0.64. (Squaring a negative number always gives you a positive number.)

If you'd been more circumspect and said there was only a 60 percent chance, then you'd have been rewarded less lavishly if you were right—your squared error would be 0.4 ^ 2, or 0.16, instead of 0.04. But if you'd been wrong, you'd have been punished less severely: you'd have 0.36 instead of 0.64.

This is the simplest form of the Brier score, for choosing between two options. If forecasters are choosing between several options, or from a continuous series of outcomes—say, the value of the pound against the dollar on December 14, 2024—then there are slightly more complicated versions. But the basic idea is the same.

After several years, Tetlock assessed the results, and it turned out that the average forecaster did very little better than random guessing. In fact, in a memorable phrase that he would come to somewhat regret, Tetlock said they did no better than "a dart-throwing chimpanzee."

He regretted it, as he wrote in *Superforecasting* three decades later, because people misunderstood it—they took it to mean that *all* experts were guessing randomly. But in fact there were distinct groups. Some thought the world was simple and could be explained (and predicted) simply—they had what Tetlock called "one big idea" that they rubber-stamped onto every situation. Others thought the world was complicated—that the specifics and details of each situation mattered, and that predictions were difficult and uncertain. He called the big-idea people "hedgehogs" and the life-is-complicated people "foxes," following a quote that Isaiah Berlin lifted from Archilochus, a Greek poet: "The fox knows many things, but the hedgehog knows one big thing." And it was the hedgehogs who did no better than the chimpanzee, if I can mix my animal metaphors.

Tetlock's example of a hedgehog was Larry Kudlow, a CNBC pundit who had previously worked as an economist for the George W. Bush administration. His "big idea," said Tetlock, was supply-side economics: he thought that tax cuts would stimulate the economy. When Bush enacted tax cuts, he expected a huge economic boost—and then said that he was right, even when the GDP and employment figures didn't back him up. Right up until the 2008 financial crisis, he continued to insist that the world was witnessing the "Bush boom." But, as Tetlock points out, this didn't hurt his career: Kudlow got a new prime-time show in 2009.

That's because hedgehogs tell nice, straightforward stories: whether that tax cuts are always good, or that we should tax billionaires more, or that the problem is that our enemies hate our freedom, or that white colonialism is the root of all evil. And those stories are easy to package for the media. So while the media isn't deliberately selecting bad forecasters, they "go looking for hedgehogs, who happen to be bad fore-casters." 27

Foxes, on the other hand, did somewhat better. Not brilliantly—lots of them were still beaten by simple algorithms such as "predict no change." But better than random guessing.

And a few did quite a lot better. The very best, the top 2 percent, Tetlock called "superforecasters."

SUPERFORECASTING, PART 2

Tetlock's work is interesting from our point of view because it found that the people who do best at forecasting the future—the superforecasters—think in Bayesian terms. Sometimes they explicitly do the calculations, but even if they don't, they think very much of priors and updates.

One superforecaster, Michael Story, who now runs the Swift Centre forecasting firm, gave me an example in an interview I did with him a few years ago for a radio documentary. "Imagine you go to a wedding," he said. "And someone asks you, do you think the marriage is going to work out? And let's assume you want to give a proper response.

"Someone who's not very experienced with thinking probabilistically might get overwhelmed with all the information in the room. You know, you can see how happy the couple are, nice music is playing, everyone's dressed nicely, there's food. And you translate the feeling that gives you into probability." You might say, "I give it 90 percent."

Forecasters call this the "inside view": judging the probability on the specifics of the situation in front of you.

But a superforecaster would start the other way. They'd look for a *reference class*, or a *base rate*. That is a body of similar events that you can use for a starting point. For instance, in this case, it might be the fact that somewhere between 35 and 40 percent of British marriages end in divorce.* That's called the "outside view": judging the probability on how often similar things have happened in the past. *After that*, they'd use the inside view to update away from their base rate. Then they might use other things—such as the age, social class, or educational level of the couple, and whether that affects the statistics; or simply the forecaster's own judgment of how well suited the couple are to each other.

It doesn't take much squinting to see that taking the outside view is simply looking for a prior probability. The prior probability of divorce is about 0.4 over the lifetime of a marriage, given no information other than they're British. Then you get more information—such as whether you think they're a well-matched couple—which acts as your likeli-

---

* It's tricky to be precise about this, because to have a firm statistic you'd need to look at marriages where at least one partner has died, so that you're looking at the entire length of the relationship. Obviously that rules out most marriages within the last forty years. But the Office for National Statistics says that about nine in every one thousand, or 0.9 percent, of couples got divorced in 2021. If you imagine most couples get married around thirty and live to around eighty, that's fifty years of marriage, so each marriage has $p = 0.991$ of surviving a given year, and $0.991 ^\wedge 50 = 0.63$. So if 2021 is a representative year, about 37 percent of marriages end in divorce. "Divorces in England and Wales: 2021," Office for National Statistics, https://www.ons.gov.uk/peoplepopulationandcommunity/birthsdeathsandmarriages/divorce/bulletins/divorcesinengland andwales/2021.

hood, or Bayes' factor, and you use that to update your prior to give you a posterior probability.

"I wasn't explicitly plugging things in to Bayes' law," David Manheim, another superforecaster, whom we've already met, says. "But it was certainly conceptually the model that I implicitly used. It's amazing how much [Tetlock's work] is a straightforward consequence of E. T. Jaynes's probability theory. How do you aggregate judgment? It's a nasty problem, but here's how the math works. How much should you care about base rates? They should be your prior. The end.

"How much should the inside view change your mind? You need a likelihood function, to say how much to shift my opinion based on this information, instead of 'What does this shiny new information say?'"

Most of us, though, don't keep base rates in our mind like that, so our beliefs are swayed by every new bit of information. "If every time you get new data you start over," says Manheim, "then obviously your estimates jump all over the place and will be way overfocused on whatever's most recent." (Aubrey Clayton would probably say that's what frequentists do.)

Of course, there's more to being a good forecaster than using base rates. For one thing, "using new data as a likelihood function" sounds nice and simple, but in most cases, you can't just do the math—yet people are still using their judgment to decide *how much* to update away from the base rate. "You have to make a value judgment on how relevant a piece of information is," says Jonathon Kitson, another superforecaster. "Not everything comes into the forecasting model, and that's where the judgment angle comes in. I'm not much of a mathematician at all, but I probably think in quite a Bayesian way, updating all the time."

There are also other tricks that good forecasters use. One is the Fermi estimate, named for the great nuclear physicist Enrico Fermi. The classic example is a problem Fermi gave his students: estimate the number of piano tuners in Chicago.

Most people would think that that's an impossible question or might just pull a number out of nowhere in particular. "I dunno, a thousand?" But Fermi broke it down into smaller, still unknown, but more easily guessable parts. Here's how Tetlock estimated it, using Fermi's system. Chicago is a pretty big city; smaller than LA, which has about 4 million people, but probably not that much. Say 2.5 million. How many people own pianos? Tetlock has a stab at one in a hundred, plus about the same number in schools and music halls, so he reckons fifty thousand pianos in Chicago.

How often do pianos need tuning? Maybe once a year, let's say. How long does it take to tune a piano? Maybe two hours. The average American works about forty hours, fifty weeks a year (says Tetlock; it sounds pretty bleak, but OK). That's two thousand hours, but let's imagine they spend 20 percent of their time traveling between jobs, so sixteen hundred hours a year. That's eight hundred pianos.

If all fifty thousand pianos need tuning every year, and each piano tuner can tune eight hundred of them in a year, then you need about $25{,}000/800 = 62.5$ piano tuners to keep all of Chicago's pianos in tune.[28] Fermi found that breaking down the estimates like this usually ended up with answers not too far from the real number. Tetlock says the real answer is something like eighty, so his estimate is impressively accurate.

You can do similar things for probability. How likely is it that—to use another example of Tetlock's—Yasser Arafat's death was caused by poison? Arafat, the leader of the Palestinian Liberation Organization, died in 2004. In 2012, Swiss researchers announced that they had discovered unusually high polonium-210 in some of his belongings. (Polonium-210 is the highly radioactive element that was used to murder the Russian dissident Alexander Litvinenko in London in 2006.) Arafat was exhumed and tests carried out to see if there was polonium in his system. And Tetlock's organization asked forecasters: "Will [the inquiries] find elevated levels of polonium in the remains of Yasser Arafat's body?"

What probability would you put on it? Tetlock warns that if you're not reflective about it, you could leap to a conclusion—"Of course Israel would do that!" or "Israel would never do that!" perhaps—and assign a probability like that, probably 100 or 95 percent in the first instance, 0 or 5 in the second. But Tetlock points to how a superforecaster did it: break down the question into parts, such as "How might polonium end up in Arafat's body and how likely is each method?" and "How long does polonium take to decay?" and "If major intelligence agencies think it's worth investigating, how likely must they think it is?"—questions you can estimate and look into individually. The superforecaster ended up giving a 65 percent probability that the investigations would find polonium—which they duly did.[29] Fermi estimates are a way of employing the law of large numbers by yourself: you make several estimates of small things instead of one big estimate, and if there's no reason why those errors should be systematically high or low, then they will tend to cancel each other out, just as Thomas Simpson found with measurements of planetary positions in 1755.

Good forecasters also use the wisdom of the crowds. That is, they update their forecasts on the basis of what others say. The average of several forecasters' predictions is likely to be more accurate than any particular forecaster's, for the same reason as the Fermi estimates—because the forecasters' errors tend to cancel one another out. But you can be more sophisticated than that. "The simplest thing would just be to average them," says Mike Story, the previously mentioned superforecaster. "Assume random noise is the reason why the experts disagree. But we also know that people differ in their ability to make accurate forecasts, and that can give you a clue for who to listen to. If they've predicted something terrible happening every six months for the last twenty years, maybe you pay a little less attention to them. But someone who's well calibrated and has a good track record, you would pay a lot more attention to." In Bayesian terms, you treat forecasts from reliable forecasters as having more information—they're like a likelihood function with a sharper peak, which moves your prior further.

And, most important, forecasters *keep score*. Note your forecasts down, publicly, and see how many of them come true, and whether your 60 percent guesses come true 60 percent of the time, and so on. It's very easy, otherwise, to forget your bad predictions and remember your good ones. "People will tell you that they want to be right about things," says Story. But that can manifest in two different ways. It could mean that they have some belief and don't want to be told that it's wrong. Or it could mean that they want to rid themselves of any beliefs that *are* wrong. "So your urge to be right can drive you in two different directions: to force your views on other people, or to throw away ideas that are causing you to be wrong.

"By being public about your forecasts, it gives you an incentive to want the information you have to be right. There's no way to force everyone to agree with you. You made a specific prediction, you wrote it down, and you wrote down your confidence. You made it public, and now there's nothing you can do. [If it's wrong,] the only way you can be right is by changing your view."

That's extremely Bayesian, once again. You have some prior belief that makes some prediction; the prediction doesn't come true; you downgrade the strength of your belief.

Perhaps this sounds very basic. But it's quite rare for people to think in terms of probabilities and percentages—we tend to fall into thinking "it will happen" or "it won't happen" (or sometimes "maybe it will happen"). And when you get people to put percentage probabilities on their beliefs, they are usually overconfident. At the forecasting workshop I went to, Mike Story and his colleagues gave us all an exercise. We were asked a series of questions, like "What year did Celine Dion have a number one with 'My Heart Will Go On?'" or "How many Premier League goals did John Barnes score for Liverpool against Leeds?" and we were told to give a range of numbers that we were 90 percent confident the true answer would fall within. (So, say, 1994:2000 or 1:8.)

"If you ask people to give you their 90 percent confidence interval," says Story, "most of the time, the numbers they give, you should have had the label of a 50 or 60 percent confidence interval." That is: their

90 percent confident answers are wrong at least 40 percent of the time. "That's standard; it's what's in the literature. Whenever I've taken a group of friends or colleagues and asked them to do it, you generally see that people are overconfident." You can test your own calibration and confidence at a number of places online—the charity 80,000 Hours has a good (if somewhat time-consuming; it's got one hundred questions) calibration exercise at 80000hours.org/calibration-training.*

As we've seen throughout this book, you can't be a Bayesian without priors. Without some sense of how probable something was *before* you saw the evidence, you can't make any claims about how likely it is *after* seeing the evidence. You can say that your COVID test is 99 percent accurate, so there's only a one-in-one-hundred chance you'd see a positive result if you actually didn't have COVID. Or you can say that you'd only see a p-value of 0.05 or lower one time in twenty if the effect you were looking for didn't actually exist. But without a prior, you can't say *how likely it is that you have COVID* or *how likely it is that you've found a real effect.*

That's true when we're making decisions in the real world, just as it is with scientific and statistical questions. What superforecasters are good at, or part of what they're good at, is finding suitable base rates— that is, priors. Sometimes, like in Mike Story's example of divorce, they're fairly obvious: you can look at the actual numbers, and see how many marriages go the distance, and use that as your starting point. But often it's more subtle. If you wanted to predict, say, the Russian

---

*I just did it and found, pleasingly, that my 80 percent guesses were correct 85 percent of the time, which isn't bad. Years of reporting on this stuff have beaten the overconfidence out of me.

invasion of Ukraine—what's your base rate? The average number of land wars in Europe per year? The average number of Russian invasions of Ukraine per year? It's a subtle art—picking an appropriate reference class to compare your example with. And then, of course, you need to know how and when to depart from that base rate.

"Choosing your base rate is the basic thing you do," says Jonny Kitson. "But I've always said that the real value of superforecasting is recognizing when the base rate is off. Land wars in Europe are rare since 1945—the annual base rate is well below 5 percent—but by December 2021 I was at about 60 percent likely that there'd be a war, and was up to 80 percent by mid-January." (You may, at this point, be reminded of the section about "hyperpriors" in chapter 3. When do you need to update your model of the world?)

Humans are good Bayesians when they're being instinctive, but most of us aren't great when we have to put actual numbers on it. Some humans, though, do it better. Even if they're not literally using Bayes' rule, they're doing the same dance.

BAYESIAN EPISTEMOLOGY

A nice thing about a Bayesian view on the world is that it dissolves a large number of philosophical conundrums that other epistemologies find extremely confusing. For instance: There's a debate going back to at least Arthur Schopenhauer, the eighteenth-century German philosopher, about definitions and identities. What is a "game," for instance? Is it something people do for fun? Well, there are lots of things that we do for fun that aren't games—skiing or reading a book. And lots of people play games for other reasons than fun, such as exercise or money.

Is it a competition, then? No: again, there are games that are co-operative or solitary, and competitions that aren't games (is the lottery a game?). Is it an activity with rules? Not all games have rules: my daughter plays great imaginative games in which she organizes her teddies into . . . honestly, I have no idea what's going on. But there are no "rules" per se.

It's not just games, of course, although that's the paradigmatic example. The United States Supreme Court once had to adjudicate over whether or not the film *The Lovers* was "hard-core pornography."[30] "I shall not today attempt further to define the kinds of material I understand to be embraced within that shorthand description," said one of the judges, "and perhaps I could never succeed in intelligibly doing so. But *I know it when I see it*, and the motion picture involved in this case is not that." How does he define what is—and is not—pornography, if there is no single defining feature or clear definition?

A huge amount of our public discourse comes down to our efforts to categorize things, groups, people, and concepts. Is [political party] fascist? Is [group] a cult? Is [person] a racist? But these categories are all fuzzy. Some things fit very obviously within them (Mussolini, clearly a fascist). Others are more controversial.

When you look at this through a Bayesian light, it's straightforward to the point of obviousness. Ludwig Wittgenstein argued that there was no one feature that distinguishes gamehood. Instead, games (and other things we struggle to define yet can clearly recognize: "I know it when I see it") have *family resemblances*.[31] The family of things we call "games" shares various features: some games have some of those features, some have others. He didn't call it Bayesian, but it clearly is.

The prior probability of any given concept being a "game" is very low. There are lots of other concepts, such as "non-Euclidian geometry" and "ennui." But if you learn that people do it for fun, and that it has rules, that's your new information, and it moves your probability up. If you find it doesn't involve competition, you move your probability down. Exactly how confident you need to be to call something a "game" is up to you (and it's not, as ever, as if you'll be explicitly doing the sums). But it's a Bayesian process.

And, of course, how you learned what defines "games" was itself Bayesian. When you were a toddler, and you first heard the word, it was probably referring to Snakes and Ladders or something like that. So your low-confidence guess is that "games" are played on flat cardboard surfaces, involve dice and snakes, and require zero skill. Later you learn that "catch" is also a game, and so is "football." You notice that these do not include flat cardboard surfaces, but that Monopoly and Clue do, so you estimate that P(flat cardboard | game) = 0.57, or thereabouts. Catch doesn't have codified rules, but all the others do. As you learn to attach the label "game" to more and more concepts, you get more accurate estimates of the probability of seeing certain characteristics in those concepts. Your prior probability that games involve balls was low, then someone pointed out that hockey, football, tennis, cricket, and Ping-Pong are all games, so you updated.

This works a lot better than some other models! Socrates defined man as "a featherless biped": Diogenes plucked a chicken and said, "Behold! I've brought you a man."[32] If Socrates had said that when he learned that something *was* a featherless biped, it vastly increased his probability estimate that that thing was a man (or a woman! Come on, Socrates, it's 411 BC, get with the times), then he would have been making a perfectly reasonable point. But there will be counterexamples, so think it possible you may be mistaken.

A comparable philosophical conundrum is the *paradox of the heap*. You have a large heap of sand. You remove one grain of sand. It's clearly still a heap. You remove another grain of sand. It's clearly still a heap. You keep removing grains, one after the other, until there is just one grain left. Now it's clearly not a heap. At what point did it stop? Which was the grain of sand that turned it from "heap" to "not heap"?

Aristotelian philosophy struggles with this. It feels pretty crazy to say that 1 million grains or whatever is a heap but that 999,999 isn't. But if you accept the premise that a heap minus one grain of sand is still a heap, then you either have to arbitrarily draw some cutoff like that, or accept that a single grain of sand counts as a heap. But with Bayesian reasoning, you don't! It's a subjective probability assessment, but you have very high confidence that a million grains of sand constitute a heap. As you remove grains, your confidence falls a tiny bit. By the time you get down to five grains you ascribe only a tiny bit of probability mass to the hypothesis "that load of sand constitutes a heap." No paradoxes or weird cutoffs are involved.

It's lovely and consistent and elegant. It avoids the Aristotelian problem of hard definitions, of clear boundaries between category X and category Y. Sometimes—most of the time—you don't have those clear boundaries, because the world is not made up of logically deductive statements. The world is not black and white.

But it also avoids the opposite problem, the Paul Feyerabend or Robert Anton-Wilson problem, that nothing is knowable at all. It admits that the world is shades of gray—but those shades of gray are *different* shades! Some of those shades are almost white, and some are almost black! I can't be certain that vaccines work, and I can't be certain that the pyramids were made by ancient aliens either. But that doesn't mean I think those two statements are equally likely to be true.

By acknowledging that beliefs and definitions are probabilistic, we can salvage the idea of knowledge and justified true belief from cod-postmodernist ideas that all beliefs are uncertain and therefore equally valid. And, of course, the beliefs that help us predict the world, that best meet the incoming information and avoid prediction error, are the ones we should have confidence in—the ones that are most likely to be "true."

CHAPTER FIVE

The Bayesian Brain

FROM PLATO TO GREGORY

We've looked at how humans are, in certain circumstances, good Bayesians: that while you can build artificial scenarios in which our reasoning goes wrong, and while we're not great at explicitly working out Bayes' rule, our decisions seem to approximate that rule pretty closely under more natural scenarios.

But we can go deeper than that. In fact, everything you perceive about the world is due to Bayes' theorem. Perception and consciousness itself is—in quite a direct sense—Bayesian.

You could reasonably argue that this is almost tautologically true. "Bayes really nicely describes the sort of problems that brains face," says Anil Seth, a neuroscientist at the University of Sussex who works on consciousness. "They're faced with ambiguous sensory information." The job of the brain is to use that information to work out what the cause of the information is. "Going from observations to what caused those observations—that's inverse reasoning, for which Bayes is very well suited," Seth says. And since I've just spent most of a book arguing that Bayes' theorem underpins all decision-making under uncertainty, and that any decision-making process is doing well insofar as it approximates Bayes and badly insofar as it doesn't, it would be surprising if our brains *weren't* approximating Bayes to some extent.

But there's a stronger claim that various scientists make, which is that Bayes' rule mathematically describes large parts of what the brain does, and that the central thing the brain does is build predictions of the world, which it then integrates with information coming in via the senses. That is, it has priors, which it updates with likelihoods, and produces posteriors. It does this at a variety of different levels—from very low-level, basic predictions about which a particular set of neurons will fire when certain muscles move, to complex, high-level, conceptual predictions like "I expect the work cafeteria will have soup today." And those predictions are tested against reality—whether the predictions match the sensory information coming in. When they don't, our brain has to update its model of the world.

This goes against what it *feels like* to perceive the world—we feel as though we see the world through a window. But we know that's not true. We know that "we" are brains sitting inside bone cavities, connected to the outside world only by fleshy strings of nerves that are linked to sensory organs. What the Bayesian brain model says is that perception is a two-way street: information travels "up" from our senses, yes, but it also travels "down" from our internal model of the universe. Our perception is the commingling of that bottom-up stream with the top-down one. The two constrain each other—if the top-down priors are strong, then it requires precise, strong evidence from the senses to overturn them.

. Scholars have wondered how we perceive the world for thousands of years. Plato's famous allegory of the cave is about perception. Prisoners are chained up in a cave, facing a wall on which the shadows of a puppet play are cast by a fire behind them. The prisoners, who have never seen anything else, think the shadows are reality, and give names to the shadows.[1] For Plato, our perception of the world is like that: we don't see reality as it is, but a shadow of it, mediated through our senses.

But Plato wasn't the first to address the question. The pre-Socratic philosopher Democritus, who lived in the fifth century BC, believed that objects in the world constantly emit tiny images of themselves, *eidôla*, made of the atoms of which the object itself is made.[2] Euclid believed that the eyes emit rays, which explore objects of the world and return to the viewer with information about those objects.[3] Those two models of perception—rays emitted from the eye, known as extramission, or physical forms emitted by objects and received by the eye, known as intromission—dominated understanding of perception, or at least visual perception, for a thousand years.

The tenth-century philosopher Abu Ali al-Hasan Ibn al-Haytham, known in the West as Alhazen, was the first to build something like a modern theory of visual perception. He argued that light was emitted from luminous objects and traveled in straight lines in all directions. That light then bounced off other objects and some of it was received by the viewer's eyes.[4]

Immanuel Kant, in the eighteenth century, said that the universe as it truly is must be unknowable, and all we ever know is the world through our senses: he made a clear distinction between *phenomena*, our perceptions of objects, and *noumena*, the things in themselves.[5] More than that, he foreshadowed the Bayesian model of the brain: he argued that our brains must have pre-embedded conceptual frameworks with which to make sense of the world, or the data coming from our senses would be a meaningless jumble. We must, in modern language, have priors.[6] We don't just passively perceive the world: we construct it, or a model of it.

This idea was taken further by the nineteenth-century German polymath Hermann von Helmholtz, inventor of the ophthalmoscope—that funny little stick with a lens on top that opticians use to look at your retina. But his great insight was that we *cannot* perceive the world as it truly is—we're not fast enough.

Our nervous system was known, at the time Helmholtz was working, to be electrical in nature. Electricity was known to travel extremely fast—the speed of light—so it was assumed that nerve signals traveled from our sense organs to our brains essentially instantaneously. Helmholtz's professor told him not to bother trying to measure it. But Helmholtz did so anyway, and discovered—to everyone's surprise— that nerve signals travel embarrassingly slowly: about 165 feet per second, or 112 miles an hour.[7] He also measured the time it took someone to respond to a sensation, such as a touch on the arm, by having them press a button as quickly as possible, and found that the time from sensation to reaction was more than a tenth of a second. This showed, he argued, that it is impossible that our perception of the world is real and immediate. It can't be, for the simple reason that the information in the world can't get to us quickly enough. If perception were direct, then we'd be constantly seeing the world a small but appreciable moment behind events. If I knocked a pen off my desk and tried to catch it, I'd be aiming for a space in the air about two inches above where it actually was.

Helmholtz argued, therefore, that our apparently effortless, instantaneous perception of the world must be an illusion. Instead, our mind makes a series of "unconscious inferences," building a 3D model of the world from the noisy, two-dimensional image projected on our retinas and the equally noisy and unclear information coming from our other senses.

He draws one example: Imagine someone holding a pen. Three of their fingers are touching the pen. But the only information each finger is sending is that of contact with a smooth cylindrical object—the direct signals from the nerves in their hand would be the same if their fingers were touching three different pens. They perceive themselves to be holding one pen because they know their fingers are close together.[8] Their model of the world shapes their perception of it.

In the 1970s, the British psychologist Richard Gregory built on Helmholtz's work. He suggested that our perceptions are essentially hypotheses—he explicitly drew an analogy with how the scientific process makes hypotheses about the world—and we test those hypotheses with our senses. He used a series of optical illusions to demonstrate his point. Optical illusions, he argued, are not just defects in our perception—they are created by the way our brains manufacture the model of the world. To create a clever illusion, we need to exploit the shortcuts our brain uses.

That's because, he said, the brain has to do a lot of work. The world as it appears on our retinas is messy: upside down and back to front, for a start (if you close your eyes and press the bottom-left of one eye, the resulting splodge of color appears in the top right of your visual field). It's also distorted by the concave shape of the back of your eyeball, and it's bumpy with the blood vessels that cover it. Worst of all, the eye is just badly designed, with the nerves from the retina pointing inward rather than out, so in order for them to get out to the brain the optic nerve has to come through the retina, leaving a big blind spot.

(Fun game. Close your left eye, and hold both your index fingers together, straight in front of you, at eye level. Keep the left finger where it is and keep staring at it, but move your right finger slowly to the right. When it's moved about eight inches, the top knuckle of your right index finger will disappear. That's your right eye's blind spot.)

"The brain's task is not to see the retinal image," wrote Gregory, "but to relate signals from the retina to objects of the external world."[9]

But there's a problem there. A signal from the outside world could be caused by a literally infinite number of things. Imagine you're outside on a dark night, and you see a single bright spot in the sky. Is it small and close—a firefly, perhaps, or a landing light on a plane? Or is it vast but far away, perhaps the planet Jupiter? Or, even vaster and even farther, the star Vega? There are two variables—size and distance—and you can explain the phenomenon "small bright light" with an infinite number of combinations of the two: closer and smaller, farther and larger, anywhere in between. "The essential problem of the brain to solve is that any given retinal image could be produced by an infinity of sizes and shapes and distances of object," Gregory wrote, "yet normally we see just one stable object."[10]

Gregory suggested that what the brain does is throw out hypotheses. Then it tests those hypotheses by comparing them to evidence from the senses. He demonstrated that when two hypotheses explain the evidence equally well, the brain can flip between those hypotheses. The "Necker cube" is the most famous example: if you're like me, you should be able to "choose" to see it as either viewed from above and to the right, or below and to the left.

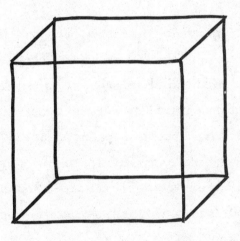

This is, you can probably see, a Bayesian model. Your hypotheses are priors. You seek new evidence from your senses to confirm or disconfirm them—that's your likelihood, your data. And you combine them to get a posterior probability distribution. In the case of the Necker cube, you have no strong reason to prefer either of your two hypotheses (the cube viewed from below or the cube viewed from above), so your prior probability is split fifty-fifty between the two, and your data fits equally well with both.

OPTICAL ILLUSIONS

In 2015, when I was working at *BuzzFeed*, one day the office suddenly became *extremely* excited. One of our colleagues in the US office—Cates Holderness, queen of finding weird stuff on the internet—had found some weird stuff on the internet.[11] It was a Tumblr post with a picture of a dress in it.

You will almost certainly remember The Dress. It was a strange moment for those of us at *BuzzFeed*, because The Dress burst out of our too-online world and became entirely mainstream. I remember going for a pub lunch with my wife and in-laws, and overhearing some total strangers walking past arguing about whether it was blue and black or white and gold. Cates's piece received over 37 million views, and basically everyone in the entire *BuzzFeed* organization, including those of us in London, was recruited to write spin-offs and reactions and so on.

The image was of a dress, with horizontal stripes. The stripes were *obviously* white and gold. But Cates's URL was "help-am-i-going-insane-its-definitely-blue." Her post included a poll that you could answer: Is it white and gold or blue and black? Of the 3.7 million people who have answered that poll over the last eight or so years, 67 percent said white and gold, while 33 percent said blue and black.

Taylor Swift tweeted that she was "confused and scared" that other people couldn't see blue and black. Justin Bieber was another member of team-blue-and-black. Katy Perry and Kim Kardashian saw it as white and gold. (I'm getting all this from the Wikipedia page.) And it was *simply impossible* for most of us to see what the other folks were talking about.

Color perception is another splendid example of Gregory's "hypothesis" hypothesis. When light hits our retinas, it can be various wavelengths and amplitudes, and there can be however many photons reaching our light-detecting cells per second. But we don't care about the wavelength of the light or the number of photons: we care about what that wavelength tells us about the object the light has bounced off. ("The brain's task is not to see the retinal image, but to relate signals from the retina to objects of the external world," remember?)

So if a spot on your retina is being bombarded by a small number of photons of relatively low amplitude, but a wide spectrum of wavelengths—that is, a dim gray color—that could be explained by one of several hypotheses. It could be a bright white object under dim light, for instance. Or a dark gray object under brighter light. Or anything in between.

This is neatly demonstrated by a famous illusion, Adelson's checkershadow illusion, created by Edward Adelson of the Massachusetts Institute of Technology. It shows a chessboard in apparent 3D, with a large cylinder in one corner. That cylinder is apparently lit from the side, casting a shadow over the board. Two squares—one in shadow, one not—are labeled A and B.

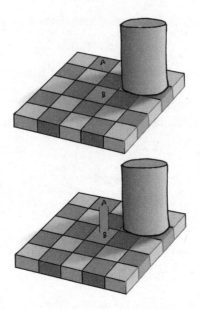

Obviously, A is one of the dark squares on the chessboard, and B is one of the light ones. But *they are the exact same shade.* You can see that when—as in the image above—we connect the two with bars of that shade. Or you can cover the page with your fingers so that only those two squares are showing. This is a pure example of your brain generating hypotheses and testing them against reality (and, in this case, being deceived by a deliberate trick). As we said, a retinal impression of a mid-gray color could come from a light square dimly lit or a darker square more brightly lit.

Receiving that retinal impression, then, your brain looks for clues for the best hypothesis. It notices that one square is in apparent shadow and one is not. So it thinks that the best hypothesis is that the shadowed square is light but dimly lit, while the other is darker and more brightly lit.

Another way you can see the hypothesis-generation in action is to look at a picture that makes no sense until you are given a hypothesis, and then you can't see it any other way. For instance, this picture of a . . . well, what is it?

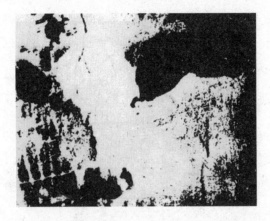

I'm not going to tell you for a few lines. Look at it and see if you can make any sense of it.

. . . I get paid by the line, you know . . .

OK. It's a picture of a cow. Its head is on the left-hand side of the image, facing you.

Can you see it now? If you can, try to see the image as a random collection of blots and splodges again. If you're anything like me, or like most people, you simply won't be able to. You have your hypothesis, you've tested it against the available evidence, and it has now snapped into place. You can't shift it.

Color perception is something that takes place—most of us probably agree—well below the level of conscious awareness. (Later on, we'll explore "high-level" and "low-level" perception in a more principled way, but for now let's just stipulate that.) And recognizing images of cows, well, "cow" is probably a higher-level concept than "gray," but it's still stuff that seems quite basic.

But how about reading?

Take a look at these two.

In the first one, did you notice that the H in the word "THE" and the A in the word "CAT" are the same shape? Whether you did or not, you certainly didn't have any difficulty reading it. Your brain knows that the the hypothesis "THE" is more likely than the hypothesis "TAE" (except in Scotland, maybe), and that "CAT" is more likely than "CHT."

And what did you read when you read the second one, in the triangle? Did it say "I LOVE PARIS IN THE SPRINGTIME"? Or did you notice that the word "THE" is repeated? Probably not. (Did you notice that I repeated it in the above paragraph, as well?)

Again, this is all Bayesianism. Your strong prior is that PARIS IN THE SPRINGTIME is more likely than PARIS IN THE THE SPRINGTIME, and that THE CAT is more likely than TAE CAT or THE CHT, so even when the evidence comes in, it's not enough to shift your posterior probability very far. You need much stronger evidence—a long, careful look—to make you realize what's actually there. In some cases, even that won't do it. When your prior probabilities are really strong, no amount of evidence can shift them. A famous example, taken from Richard Gregory, is the "hollow mask" illusion. Look at these images of a mask of Charlie Chaplin: [12]

The first image is of the mask facing us. But then it rotates. The fourth image shows it facing away: the face is concave, hollow. But our minds have a very, very strong prior that faces point outward, not inward, so we see it as convex. This prior is *so* strong that even when we look closely, we can't make ourselves see it differently. (If you Google it and find videos, it's even more bizarre. As the mask turns around, your brain *flips* somehow: even though you've just seen the mask turning away, you can't see it as convex.)

Hopefully all this makes what was going on with The Dress much clearer. The information entering people's eyes was all pretty much the same—a selection of photons at certain wavelengths and amplitudes. But that information was compatible with (at least) two plausible hypotheses: a dark blue and black dress under bright, yellowish lighting, or a white and gold dress under dimmer, blue-tinted lighting.

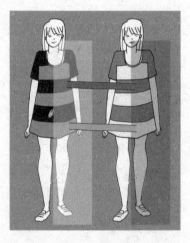

What's interesting about The Dress is that most people are unable to pop back and forth between those two hypotheses, as with the Necker cube, and once the "real" color of the dress was revealed, it didn't then snap into place as with the picture of the cow. (The dress was actually blue and black: the woman wearing it shared another picture.) One possibility is that people start with different priors: one paper suggested that "morning people" might have a prior that the light would be bluer. (It found, at best, inconclusive evidence for that hypothesis, though.)[13]

The exact mechanisms behind our differing perceptions of The Dress are still not known. But the idea is familiar: our prior probabilities, our top-down model of the world, informed our perception of incoming information. The Dress was a Bayesian phenomenon.

REALITY AS CONTROLLED HALLUCINATION

A guy called Richard Fitzhugh once tried an interesting experiment. Looking *only* at the nerve impulses reaching a cat's brain from its retina, could he determine what that cat was seeing? (I assume this involved a certain amount of cutting up of the cat.)

The information reaching your (or a cat's) brain, we sometimes forget, is energy. A photon hits a receptor cell and causes a tiny chemical change, which sends a chain reaction along a nerve. Pressure on a fingertip does something similar. What our brains receive is a series of energetic spikes from the nerves that lead into it. When nothing particularly interesting is happening to some sensor, it is largely quiet, firing somewhat randomly a few times a second. But when something changes—for instance, when a light flashes brightly—the nerves fire more concertedly and a greater number of signals reach the brain. Fitzhugh tried to develop a statistical method for determining when the cat had seen a flash just by the signals passing through the optic nerve.[14]

He was successful (he learned, when he checked his results against reality). But actual brains must do something much more difficult: they have to determine not just between "flash" and "no flash" but "dog," "mouse," "car," "owner," "bowl of Whiskas," "attractive cat of opposite sex," and an infinitude of other possibilities. (I'm still referring mainly to cat brains here.) All they receive is different frequencies of electrochemical energy spikes from different inputs. Somehow, they turn that into a fully realized world of physical objects and social interactions.

We've seen that there's something going on in our brains that involves priors, predictions, hypotheses. But now let's flesh it out a bit. The neuroscientist Chris Frith says that our perception of reality is a *controlled hallucination*.

Imagine that I'm looking at a coffee cup on my desk. (For some reason, every book I've read about this talks about coffee cups. I wondered if they were all cribbing off each other, but I've decided it's because they're all written by people who are sitting at a desk, scanning their eyes around for an example, and lighting upon the same one. So I'm going to go with tradition.) The intuitive model of perception says that I perceive the coffee cup through bottom-up signals—that is, signals come in through my eyes, like a television camera transmitting pixels to a TV screen in our brain: signals of basic features of reality like color, lines, shape. Lower-level processing in our brain takes those features and builds them into ever more complicated ideas, which are then compared against memories and knowledge of the world and assigned labels like "mug" and "coffee."

That model of bottom-up perception drove a lot of cognitive science for many years. But now the understanding is that it goes something like this.[15]

Instead of our image of the world coming in from our senses, our brains are making it up, constantly. We build a 3D model around ourselves. We're predicting—hallucinating—the world. There's not just a bottom-up stream of information—there is, vitally, a top-down one, as well. Higher-level processing in our brain sends a signal *down*, toward our nerve receptors, telling them what signals to expect.

So when I look around my desk, and my eyes move to a certain point, higher-level parts of my brain send signals to lower-level parts, saying things like "Expect a pink coffee cup next to the keyboard." Those concepts are broken down by the lower-level processors into things like "A squat pale cylinder shape roughly thirty degrees of arc from the center of my visual field." That in turn is broken down into more basic concepts: this color *here*, a vertical line *here*, and so on. And those are translated into the maximally basic, machine-code-level version that Fitzhugh was dealing with: expect *these* axons in the optic nerve to fire roughly *this* many times per second. They're guesses, predictions, hypotheses, cascading down from the conceptually complex higher levels to the utterly minimalist nerve-signal levels.

At the same time, signals are coming up from those nerves: *these* nerves are firing *this* many times. The cup is where I expected it to be, so the nerve signals match the predicted patterns. Because there's nothing unexpected there, my model of the world is left unchanged. The coffee cup is where the coffee cup should be, so there's no need to send any signals further up the chain. The hallucinated scene around me can stay in place.

But now imagine that I reach for that coffee cup. I *believe* it to be full of hot coffee. I move my hand to where I expect it to be and grasp it (higher-level expectations of a hot cup of coffee; lower-level expectations of a cylinder of certain weight and temperature; machine-code expectations of proprioceptive and temperature-sensing nerves signaling just *so*). But when my hand closes around the mug and starts to lift it, the nerve signal patterns don't match expectations.

Now things start to happen. When the predicted pattern doesn't match the received one, the low-level processor bumps the problem one place higher up the chain. If the slightly higher-level processor can explain it, then it will do so, and will send new signals back down the chain. If it can't, it sends its signal higher up, until it reaches the high-level areas that can explain it with the conceptually complex understanding that I finished my coffee a quarter of an hour ago, the cup has long since gone cold, and I need to go and boil the kettle if I want another.

What's important then is not the signals coming up from my nerves per se, but the difference between those signals coming up from my nerves and the *predictions* cascading down from my higher-level brain regions. The crucial phrase is *prediction error*, that difference between expectations and results. In this framework, your brain is constantly trying to minimize prediction error, making its model as close to reality as possible by updating it as new signals come in.

Statisticians and machine-learning people might recognize this as equivalent to a "Kalman filter"—an algorithm that takes various measurements, uses them to estimate some unknown quantity that you want to know, and then uses that estimate to make predictions. For instance, your phone's GPS receives signals from various satellites, uses them to estimate your position, and then uses that estimate to make predictions of when it will receive the next signals, and the whole thing begins again. Prior, data, posterior, which then forms the new prior.

The brain has to do a lot more than that, of course. As well as predicting what signals it will see, your brain is having to predict what the signals it is sending to your muscles will do, and how they'll affect the signals from your senses, in this fantastically complicated dance—signals coming up and down (and across), "handshaking" with one another, the various processing regions checking predictions against findings; the confidence and specificity of the prediction and the precision of the incoming information are all weighed and judged. And, of course, your brain is taking information from various modalities—vision, hearing, touch and smell and taste, but also your internal senses of where your body is and how it's arranged and whether you're hungry or thirsty or horny or whatever—and combining them.

But fundamentally—and you can probably see it by now, and you're probably bored of me saying it—it's *another Bayesian system*. Your predictions are the priors, the sense-data is the likelihood, and the updated predictions are your posterior probability. And, crucially, what you *experience* is not the data from your senses but your predictions—predictions constantly updated by information from the senses, yes, but the world you live in is the prediction, not the data. "What we experience is best described as a Bayesian inference about the *causes* of sensory data," says Anil Seth, the neuroscientist.

I thought I'd be going too far to say that our conscious experience basically *is* our Bayesian priors. But both Seth and Frith cheerfully agree with me. "Consciousness is our model of the world, not the world," says Frith. "The content of our perception is the content of these top-down predictions," says Seth. So: consciousness is Bayesian.

DOPAMINE AND FANCY COMPUTERIZED ROBOTS

Once again, I need to be a bit careful. It's easy to say that anything is Bayesian if you're not actually doing the math. "You have a guess, and then you get some new information, and you change your guess! Totally Bayesian!" Perhaps you might want a bit more convincing.

Here's one thing. As I mentioned, our brain has a particular challenge that we haven't really discussed so far—how to integrate information from different senses. If I'm talking to someone, I can use the information from my ears—the sound of their voice—and from my eyes, by looking at her mouth as she talks.

What a good Bayesian would do is put more weight on whichever sense gives the most precise information. Chris Frith and Anil Seth both mentioned one experiment that showed that this is exactly what happens.[16] "It's a beautiful experiment," said Frith. "Not by me. You bring together vision and touch."

The experimental subjects were asked to estimate the width of a ridge on a board, using their eyes and hands. But the board and ridge weren't real: they were projected onto a mirror from a screen above, with the subject's hands underneath the mirror attached to what Frith called "fancy computerized robots" and the paper's authors called "force-feedback devices." This allowed the researchers to increase or reduce the precision of the two inputs as much as they liked, by adding static to the image or imprecision to the feedback from the fancy computerized robots.

Under normal circumstances, vision is a more precise sense than touch, so the subjects' estimates were much more based on their visual sense than the "haptic" (touch) feedback. But as the experimenters added noise to the visual field, the more subjects were influenced by touch.

The interesting point, though, was that the experimenters *also* modeled how an observer would integrate the information from the two noisy senses if they were being perfectly Bayesian, using "maximum likelihood estimation," which you may remember as the thing that Ronald Fisher and Karl Pearson fell out over. As the standard error on the input from each sense increases—that is, as the curve on the graph gets flatter and wider—the amount it should influence our beliefs becomes less, in the way Bayes predicted.

What the experiment found was that how humans *actually* integrated the information was extremely close to how an ideal Bayesian observer would. Our brains are taking noisy data and using it in a close-to-Bayes-optimal way.

You can see this integration (although I guess you'll have to take my word for it that it's Bayes-optimal) in a number of audiovisual illusions online. The most famous is probably the McGurk effect.[17] If you watch the video, you'll see a facial close-up of a man apparently saying "bah . . . bah . . . bah" and then "vah . . . vah . . . vah" over and over again. But the sound is the same throughout—always the same "bah." The only difference is that the man in the video's lips are pressed together when you hear "bah," while he puts his top teeth over his lower lip when you hear "vah" sounds. Your brain takes the (extremely precise) information from your eyes and overrides the somewhat less precise information from your ears.

There are lots of audio illusions like this. One that I find *incredibly* difficult to understand is that the same ambiguous noises will be heard as either "green needle" or "brainstorm" depending on which words you're reading at the time.[18]

There's more to it, as well. When we are expecting something, our brains react to the *prediction* of that thing more than the thing itself. Frith also pointed me toward a 2001 paper by the neuroscientist Wolfram Schultz, which put electrodes in monkeys' brains (which I agree is not an especially nice thing to do) and looked at when dopamine-releasing cells were active.[19] (I'm going to sidestep a massive bun-fight here about whether it's OK to call dopamine "the reward chemical" or "the pleasure chemical." It is a neurotransmitter, it has many roles, it's not as straightforward as saying it's how our brain tells us we're happy or whatever. But it has a relationship with reward.)

In this paper, the experimenters taught the monkeys to expect a reward—a gush of tasty fruit juice—after they saw a bright light flash. The monkey would be shown the light, and then a second later be given a gush of juice directly into its mouth.

Fans of vivisection will be reminded of Pavlov and his dogs, learning to associate him ringing a bell with him bringing food. Pavlov noted that, in time, the dogs started to salivate when the bell rang, not just when they saw the food. Similarly, Schultz's results found that, at first, there was a spike in the dopamine cells' activity just after the juice arrived—the monkeys were responding to reward. But as time went on, the spike started to come just *before* the juice—the "reward" came with the flash. The actual arrival of the juice itself caused no more activity.

It got more interesting still. If a squirt of juice came along *without* a flash of light first, the dopamine cells' activity would spike at the unexpected reward, as before. But if the light flashed, but then the juice didn't arrive, the dopamine cells' activity *dropped*, below baseline levels. The reward was expected, but failed to materialize, and the monkey (or at least the monkey's dopamine-producing cells) was disappointed:

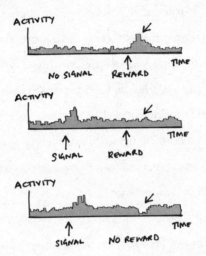

This is the lowest-level version of the "your brain is a prediction engine" model. At this very basic, machine-code level, your senses predict the world, and when the world is as they predicted it, they don't send any more signals. But when the predictions are *wrong*, they send signals higher up.

This is crucial. At every level in the hierarchy, what we experience is *what we predict.* Those predictions are checked against reality. If reality agrees, fine. If it doesn't, we have a prediction error, and *then* signals are sent further up.

Other studies[20] have found that prediction error, rather than fulfilled predictions, drives nerve signals. For instance, in the retinal ganglion, a bundle of nerves in the eye, cells "signal not the raw visual image but the departures from the predictable structure."[21] And even at this very low level, there seems to be a Bayesian relationship. New information is integrated and becomes part of future predictions in something like a Bayes-optimal fashion.[22]

We can flesh out the picture we built in the last section a bit more now. The crucial bit is that the more precise the predictions, and the more precise the data from your senses, the more attention your brain pays to them. All the time, signals are coming down from your high-level processors telling your lower-level processors what to expect. They translate the signals into lower-level predictions again, and check them against the sense-data coming up from even lower-level regions. At each level, the information from the level above acts as the "prediction" and the info from the layer below acts as the "data."

But it's all probabilistic. Some perceptions and predictions are more confident than others. The more confident predictions are usually called "high-precision." A perception of a cow standing in a field right in front of you, ten feet away on a clear sunny day, is high precision. A glimpse of a dark shape seen through murky water while you're snorkeling is very low precision. A prediction that a hammer will fall down when you drop it is high precision; a prediction that inflation will be below 5 percent next year is very low precision.

At each layer, something like this is going on. It has its prediction from above, and its sense-data from below. It uses Bayes' theorem to put them together. If the two roughly match, then the prediction is about right. (In more formal Bayesian terms, if the likelihood data is close to the prior, then the posterior probability won't be very different either.) In that case, the layer doesn't send much in the way of signals up or down—it just sort of goes, "OK, then, seven of the clock and all's well."

But if they disagree, it can go one of various ways. One, let's imagine the sense-data is very low-precision, and it disagrees with a very high-precision prediction. You are walking across Hampstead Heath on a misty day and you see, out of the corner of your eye, a hundred yards away through the fog, something roughly the size and shape of a Cape buffalo. Your brain has a very confident prediction that there are no Cape buffaloes in North London, and the data from your eyes is messy and imprecise, so the predictions override the sense-data. Imagine data with a very different mean to the prior, but a huge standard deviation—a wide, flat curve on the graph, which doesn't move the tall, narrow curve of the precise prior much. The layer again stays quiet, with little signal sent higher up.

Two, higher-precision sense-data disagree with the prediction, and the new information does move the needle: the prediction is probably wrong, according to the Bayesian equation. In that case, the layer has a prediction error, or "surprisal." It will therefore fire its neurons, alerting the layer above it in the hierarchy. The worse the mismatch, the more strongly the neurons fire. Extremely precise predictions being contradicted by extremely precise sense-data—the clouds clear a bit, the sun comes out, and you look over toward Parliament Hill and what the hell that actually is a Cape buffalo *Jesus Christ*—send a very strong alarm to higher levels.

And when that higher level receives the alarm, the information acts as its sense-data, and the whole thing goes again. The higher-level processor tries to see if it can make sense of the information with higher-level models of the world. If it can, then it gives new predictions to the lower levels to make it all fit together, and doesn't alert the higher levels; if it can't, it sends another alarm even higher up. At each level, the processors are reconciling bottom-up data with top-down predictions, and either using them to make new predictions, which they send back down the chain, or raising alarms higher up the chain when they can't make it make sense. The greater the mismatch—the more the prior is shifted—the louder the gain, or "volume," of the signal sent upward.

The key thing is that the brain hates prediction error. It wants to minimize the difference between its predictions and its sense-data. It really wants its predictions to be right. So it calls attention to mismatches so they can be sorted out. Under this model, "attention" is literally just when the higher-level processors and high-precision sense-data are focused on some aspect of your environment. Something *grabs* your attention when the bottom-up data about it coming from your senses doesn't match the top-down prediction coming from your brain—sending a loud, urgent signal right up to the top.

TENNIS, WORDLE, SACCADES

So far I've talked about perception as though it's something that just happens to us, as though we sit like a sea sponge, absorbing information from the outside. To be fair, it's easier to talk about perception like that, and when you're trying to build a basic model, that's probably the way to do it. But we don't behave like that.

As well as simply absorbing information, we seek it out. We move around in the world. We move our heads closer to things, or get up and walk over to something to check it, or pick it up and put it in our mouths. We might get a telescope to check whether that light in the sky is a planet or a star.

This gives the predictive brain two new challenges. First, predicting what the effects of its own movements will be, and second, predicting what the best move is to make in order to gain the most information possible about the world.

The Bayesian model of how this works is called *predictive processing*, and its founding father is a neuroscientist called Karl Friston, who works at the National Hospital for Neurology and Neurosurgery in Queens Square, London.

"People were fluently talking about the Bayesian brain hypothesis by about 1990," he told me. "But it was really hijacked by the sense-making side, by the perception side. People forgot about motor control, decision-making, the action side of things—the way you actually go and gather your data. That brought a very much broader problem to the table."

Once again, we tend to think of perception and action as separate things. We see the world with our senses, then we decide what to do, and then we do it. As we've already established, though, we *don't* see the world, exactly. We predict the world, and update those predictions in a Bayesian way through new information.

The trouble is that the signals we receive from the world—the patterns in which those nerve cells or the dopamine-producing cells and so on fire—are dependent not just on changes in the world, but on changes in our bodies. If a horizontal line of retinal cells fire in sequence, that could be because a bright light has moved from right to left in front of me. Or it could be because I've turned my head, causing a stationary light to move across my visual field. So our brains not only need to predict the signals coming from the world—they also need to predict how the signals coming from the world would change, if we performed some action. Then they need to subtract those predictions from the predictions of how the world itself is changing, to give the impression of a stable reality.

But there's more to it than this. The brain wants to reduce prediction error, as we've seen. It can do that by changing its beliefs to match the world—I believed there was hot coffee in my mug, I grabbed the mug, it was cold, I no longer believe that. But it can also change the world in order to match the beliefs. In that example, perhaps, you might go and refill the mug with hot coffee. In the end, for Friston at least, we can rephrase *all* our mental activities—even our desires and decisions—in the same terms as predictions.

That's for later, though. For now, we'll start with the simpler stuff.

First, there's a direct sense in which action requires prediction. If you want to move your arm, your brain has to predict which sequence of nerve firings will perform that action. Or, to look at it another way, when your brain fires a particular pattern of nerves, it has to predict what action your body will make.

These are two separate things, and, according to one model of action at least, it turns out your mind does both. The first is called the *inverse model*, and the second the *forward model*. "The inverse model is what signals I have to send to my muscles," says Frith. "It's problematic, because you have a goal—I want to reach and grasp something—but there are an infinite number of ways I could do it.

"Meanwhile, the forward model is fixed. Given the signals you decide to send, you can calculate exactly what will happen." The two models, says Frith, work in parallel—your brain runs both simulations simultaneously and checks them against each other. So if you have some goal (pick up a coffee cup), your brain predicts what sequence of nerve firings would do best at that, and at the same time it takes the predicted sequence, predicts what would happen if you do it, and sees if the two models match: "Does this inverse model actually result in this goal that I'm aiming for?"

This means, incidentally, that we can learn by imagining. We can imagine ourselves doing some goal—right now, I'm imagining myself kicking a soccer ball with the side of my foot—and by predicting which sequences of nerves would achieve that goal, and then predicting what would happen if we fired that sequence of nerves, we can genuinely improve at some task without actually doing it outside our imagination.

But what's also important is that our brain has to predict what sensations we will experience if we make some move. If you're running to catch a bus, the bus will be growing larger, and bouncing up and down, in your visual field. But your perception of the bus will be of a stable object of unchanging size, because your brain has predicted the effect that the signals it sends to your muscles will have on the signals it receives from your eyes.

Your brain then has to subtract those movements from the movements it expects about the world. (If the bus actually is moving toward you as you're running, then you want to be able to see that the bus is moving.)

And your brain also performs actions that are not intended to complete any given task themselves, but are meant to gain information about the world.

Here's an analogy. You've probably played Wordle, since everyone in the world has. If you haven't, it's a game where you have to guess a five-letter word. You have six goes, and each time you have to enter one valid (US English) word. If any of the letters in your guess are exactly right—the right letter in the right place—that letter will go green. If the letter is in the word, but not in the right place, it'll go yellow.

There are about two thousand words in Wordle's database, so your probability of guessing right on the first go is about one divided by two thousand, or p = 0.0005. You could just try to guess six random words, but you'd only have about a 0.3 percent chance of getting it right.

So instead, here's what I do: I try to gather information. I might put in a word with lots of common letters, such as ARISE. Let's say it comes back with two hits:

How many words are there in the Wordle database that have an A somewhere in them and an E on the end? I don't know exactly, but a few dozen perhaps. Suddenly the probability I can ascribe to different words has changed enormously. Instead of putting p = 0.0005 on all two thousand words equally, I can put maybe 2 percent likelihood on, say, "PLACE," or "LEAVE," but 0 percent on "BRACE" or "GLEAM."

What do I do next? If there are, say, fifty words left, then I could start guessing, but still I'd probably only have a 10 percent chance of getting it right (five guesses, fifty options). So I might want to narrow it down further.

If so, then some words are going to narrow it down more than others. I could put RAISE in. But since I've already tried all those letters, the only bit of information I could get out of it is whether the A is in the second place or not. (And it probably isn't: I bet there are more remaining words with the A in the middle, like GLAZE or CRAZE, than in the second place, like LANCE or MANGE.)

The most commonly used letters in the English language are probably E, T, A, O, I, N, S, H, R, D, L, U. So you could go for another word full of those common letters. I often go for DONUT. Or you might go for something with only one vowel because you figure there are two already in there.

(I hope, by the way, that I don't have to say, "Look, it's Bayesian!" again by this point. Your prior probability of it being any given word is 1/2,000; then you get your new data, and you update to a posterior probability, precisely using Bayes' rule.)

The point is that there are moves you can make that you know won't actually fulfill the task—it's definitely not DONUT, there's no A or E in there—but that provide you with information to do the task. Some moves are better than others. One of them (at least) is the *Bayes-optimal* move, the guess you can make that will reduce your search space the most.

I ended up guessing BOTHY and then CHAFE, for the record, which was correct, but I think I probably got a bit lucky.

This idea of "Bayes-optimal design" goes back to Dennis Lindley, says Friston. "If I had, now, to choose what data point to gather next," he says, "where to look next, what would be the best query or question?"

And that idea has become central to what Friston, Seth, and others talk about when they talk about perception. The brain is not only passively perceiving, but actively seeking out information to reduce its uncertainty in the world. "You can frame it," says Seth, "in terms of actions that are instrumental, to reach the desired goal now, or epistemic actions that maximize the information gain."

A marvelous example of this is saccades. As we've talked about, although it *seems* like you see the whole of your visual field in glorious, colorful detail, that's not the case. Only the center of your retina, the fovea, can resolve images sharply or see color. The rest is filled in—predicted—by your brain. (If you choose a random card from a deck without looking at it and hold it out to the side and behind you, then move it slowly into your visual field, you won't at first be able to tell whether it's red or black.)

In order to fill in this detail, your brain moves the fovea around. When your eyes move from one point to another, they do so in a rapid move called a *saccade* (pronounced sack-ARD). They're so fast that, to other human eyes, the pupil seems to jump—you can't see the movement, just a change of position.

But where do our eyes saccade to? One possibility would be that they jump to the most salient point in the visual field—the brightest or most standout objects, a red dot in a mass of green dots, or a single vertical line among a load of horizontal ones. That would be a bottom-up model of perception, where the details of the scene drive what we look at, and how we build our understanding of the world.

That's not what we do, though. Instead, through clever experiments, researchers have shown that our eyes saccade to *where we expect the action to be*. If you track someone's eyes during a game of, say, tennis, the eyes don't jump to where interesting things are, but where the person anticipates they will be. "Saccades are launched to regions where the ball will arrive in the near future," says one paper.[23] "Crucially, at the time that the target location is fixated, there is nothing that visually distinguishes this location from the surrounding background of the scene."

Looking ahead like this to currently undifferentiated but soon-to-be important places in the visual field allows the brain to reduce uncertainty as much as possible. In tennis, for instance, the ball moves far too fast for our eyes to track it smoothly. Instead, the brain predicts the most important, information-dense points on the ball's journey—when receiving serve, say, the point of contact with your opponent's racquet, the point at which it bounces, and the point at which it meets your racquet. The science-y tennis blog *Fault Tolerant Tennis* describes it like this: "Multiple times throughout a fast-moving ball's flight, you'll execute the same visual pattern: Predict a future location of the ball. Perform a saccade, fixating there before the ball arrives. Hold your gaze on said point until the ball arrives. Track the ball briefly through your focus. Repeat."[24]

Because the ball is moving through the foveal region of your visual field at these critical moments, your brain is able to get the maximum possible amount of information about its flight. If your prediction is wrong, it will be extremely obvious. If it's right, you'll get lots of high-quality information about the ball's ongoing movement as it passes through your focus, allowing you to predict where it will be at the next critical point (where your eyes will saccade to once the ball leaves your focus).

Of course, this means that perception is in fact a highly skilled operation. I'm a soccer fan, but I'm also rubbish at soccer; I never played as a child, and so have all the grace and fluency of Treebeard the Ent. But I also notice that it means I'm not as good a soccer *watcher* as others. I can't predict the body positions and the contacts between foot and ball as well as my friends can, for instance—they seem much more able than I am to tell when someone caught the ball beautifully on their laces and when they awkwardly toe-punted it, presumably from years of playing and seeing how contact on the laces or toes relates to subtle variations in body position or how the ball comes off the foot.

That's visible in research, as well. Learner drivers' eyes tend to focus on the road just in front of them, while seasoned drivers look farther ahead for important details like junctions and hazards.[25] Cricket and tennis players get better at predicting where the ball will bounce. Novices aren't good at predicting where the action will be, so they have to make very imprecise predictions, while experts have a well-constructed model that allows them to gather highly precise information about the world. Just as a good Wordle player has to make good judgments about which words will give them the information required to guess the answer, a human brain has to know where to seek information to best carry on building its Bayesian model of the world.

HOW COME SCHIZOPHRENICS CAN TICKLE THEMSELVES?

Why can't you tickle yourself?

Sorry. Let me rephrase that. *Can* you tickle yourself? I probably ought to ask, because if you can, that might be important. Most people can't. But, it seems, a subset of people can.

In 2000, a paper was published in the journal *Neuroreport*,[26] by the neuroscientists Chris Frith, Sarah-Jayne Blakemore, and Daniel Wolpert.* It made a surprising prediction, and tested it. The prediction was that people who suffer from schizophrenia *can* tickle themselves.

The reason they predicted that has to do with Bayes' theorem.

We've seen that our experience of the world is actually our prediction of the world—our Bayesian prior—rather than the content of our senses, although it is constrained by the data from our senses. A crucial part of how that works is that we pay less attention to sense-data that we can accurately predict. Remember, if you're a being that moves around

---

*A self-indulgent aside, but I feel there's something rather Bayesian to note here. Chris Frith's wife Uta, Sarah-Jayne Blakemore's father Colin, and Daniel Wolpert's father Lewis are or were all giants of their various fields: respectively, psychology, neurobiology, and developmental biology. What are the chances that the three authors of one paper would all have famous scientist relatives? If we use the base rate of famous scientists in the population, astronomically low, but if we remember that jobs often run in families, somewhat higher. Still, pretty remarkable, I think.

in the world, sometimes changes in your sense-data will be caused by changes in the outside world, and sometimes they'll be caused by your own movements. You need to be able to tell the two apart, and discount the latter, so that you get a sense of a stable world that you can detect movement in. (When you run or walk, you don't get the sense that the world is bouncing, even though all your sense-data is compatible with the hypothesis that it is.) The highly predictable signals are subtracted from your sense of the world. "When you move," Frith says, "the movements you cause are suppressed, leaving the movements that you have not caused, which are usually more important."

That's also why, incidentally, we tune out background hums, and why we suddenly notice them if they stop; also why if there's, say, a repetitive piece of music that plays over and over again, with the same four-beats-to-the-bar rhythm, and then after twenty minutes it misses a single beat, you'll hear the absence almost as a positive noise. The background or repetitive sounds are highly predictable and so your brain predicts them and starts ignoring them. If they stop unexpectedly, that's *not* predicted, so it's very obvious.

Anyway. This is true in all our senses. There was a neat experiment that shows it with touch.[27] People were arranged in pairs and asked to rest their left index finger on a board. On top of the board was a device that pressed down onto that finger, controlled by the other player. The two players would take it in turns to press a button. The harder they pressed the button, the harder the device pushed down on their opponent's finger. The task was to match the force the other player had used.

Each time, players overestimated how much force their opponent used, meaning that the pressure used escalated each turn. They *also* looked at what happened when a machine pressed down on someone's finger, and they were asked to match the force used by pressing on their *own* finger with the device. Again, they consistently overestimated the force required. (The authors speculated that this mechanism explains why children's playground fights tend to escalate—each child honestly believes that they are only hitting as hard as they were hit.)

But when people were asked to do the same using a joystick to control the device instead of a button, so the force was harder to predict, people got better at correctly judging how much force they were using. That's consistent with the idea that we discount strongly predicted sensations: we just don't feel them as much.

So, tickling. The same applies. If you try to tickle yourself, your brain can predict the sensations it's going to receive, with very high precision. If you were to stroke my palm and record my brain activity while you did so, you'd see a sudden spike in the number of neurons firing in the relevant bit of my cortex.[28] But if I were to stroke it myself, there'd be very little increase. "When you touch yourself," says Frith in his book,[29] entirely deadpan, "your brain suppresses your response."

Here's an interesting thing. People with schizophrenia are less susceptible to many optical illusions than the average person. The "hollow mask" illusion, for instance, can be used as a diagnostic tool—one study found that about 30 percent of schizophrenic people see through the illusion, compared to 10 percent of the general population.[30] If you're a medic dealing with a difficult-to-diagnose case that might or might not be schizophrenia, it's worth checking if your patient sees the hollow mask as convex or concave.

What appears to be going on is that schizophrenics have weaker priors than we do. Their predictions of the world are less precise, so they can—for instance—correctly assess a backward mask as hollow when the sense-data fits that hypothesis.

Unfortunately, that has other, less beneficial effects. For instance, schizophrenics often report that their body is under the control of some outside force—that when their arm moves, it's not them who's moving it. Frith tells a story in his book of a patient called PH. "My fingers pick up the pen," she says, "but I don't control them. What they do is nothing to do with me."[31]

The Bayesian explanation is that PH's predictions of how her arm will move are less precise, so that when she moves her arm, that movement is not "subtracted" from her experience in the same way it would be for a neurotypical person's. She experiences the movement unsuppressed, just as if someone elsse were to pick her arm up and move it for her.

It also explains visual and auditory hallucinations. Schizophrenic people often report hearing voices in their head: "thought insertion." But under this model, they're just hearing the voice all of us, or at least most of us, hear—our own internal monologue.* The difference is just that, for most of us, those voices are predicted, and therefore, the sensation is suppressed, like with the moving arm. But for schizophrenics, it's as shocking and as loud as if someone spoke inside their mind.

And small visual disturbances that generate minor low-level prediction errors might get explained away by higher-level processors in neurotypical people, because their priors are strong enough to say, "Come

---

*Apparently we don't all actually have an internal monologue, which I find very strange.

on, faces don't point inward," or whatever. They would predict visual changes from moving their head, or noisy data coming in from their blotchy retinas, and suppress it in the usual way.

But schizophrenic people, with their weaker priors, don't predict the world so precisely, so that same data coming in causes prediction errors, raises alarms, and gets incorporated into their model of the world. And because the errors are random—they're not generated by real things in the world but by noise in the sense-data or by unpredicted movements—the brain has to come up with bizarre hypotheses to explain them. Perhaps the pulsing of blood through the veins in our retinas makes rhythmic changes in the sense-data we all receive, but most people predict it and suppress it. Schizophrenics, though, might have to explain it as "The walls are breathing."

I've been talking about relatively low-level predictions here, but the same seems to apply to higher-level concepts—schizophrenic people might get inappropriately surprised that someone with their first name is mentioned in the newspaper, or that they saw a car with a license plate that has the number thirteen in it, or whatever. Because it's caused a prediction error, it has to be explained away with hypotheses, and that creates delusions, such as that the TV or newspapers are giving them secret messages.

You may now be able to see why I brought up tickling. Most of us can't tickle ourselves because we can predict the sense-data that we will receive so accurately—a finger tickling us *here* at *this* moment, another one *here*—and those predictions are subtracted from our experiences. But schizophrenic people apparently don't have such precise predictions of that data. So—Frith, Blakemore, and Wolpert hypothesized—

they should be able to tickle themselves. Or to be more specific, they predicted that people who experience auditory hallucinations and other symptoms of schizophrenia are more likely to say that stroking their own palm gives just as "intense, tickly, and pleasant" a sensation as when someone else strokes it.

And that's exactly what they found. People with symptoms of schizophrenia were just as susceptible to tickling themselves as they were to other people tickling them.

"What I love about it," says Anil Seth, "is it's a very unexpected prediction. Who'd have thought that this would be characteristic of schizophrenia? A Freudian wouldn't come up with this hypothesis. The only way to come up with it is by thinking about the brain in this [Bayesian] way. And for me that's the merit of a good theory—it makes predictions other theories wouldn't predict. Like relativity. My personal bugbear is people coming up with theories that are consistent with everything. Make it predict things! The schizophrenia thing nails it."

HAVE YOU EVER, LIKE, REALLY LOOKED AT YOUR HAND, MAN?

This is more speculative, but there's a growing body of thought that believes you can talk about depression in Bayesian terms. More than that, some scientists think that you can treat various psychiatric conditions, including depression, with psychedelic drugs like magic mushrooms, and that they work in a Bayesian way.

I don't want to put too much weight on this. I think there's plenty of evidence that the brain is Bayesian, and if it turns out (as it may well, still) that psychedelics are not effective antidepressants, then that won't undermine the general point. But it's a lovely, neat hypothesis and has some preliminary evidence for it, so let's go through it.

First, there's some evidence showing that psilocybin—the active ingredient in magic mushrooms—reduces depression. There was a 2021 paper[32] that found that psilocybin was as effective as escitalopram, one of the most effective existing antidepressants. Now, we should be careful: it was a small trial, and (for obvious reasons) it's quite difficult to do a "blind" trial of psychedelic drugs. A "double blind" trial is when neither the patients nor the administering researchers know who gets the treatment and who gets the control. It's meant to reduce the impact of the placebo effect. But if you suddenly start hallucinating, you'll probably have some idea that you got the good stuff. The researchers included a clever trick, which was to give the control group a tiny dose of psilocybin, too small to have an effect, in the hope that it would leave people with some uncertainty. But it probably wouldn't have confused people very much.*

At least four other studies[33] have found similar results, but (1) they all suffer from the exact same problem ("I'm pretty sure I'm not on the placebo, Doctor; you appear to have turned into a camel"); and (2) the slight trouble with any studies in this area, like studies into, say,

---

*I should also point out that the psilocybin was administered under careful laboratory conditions, under medical supervision, and alongside therapy, to people with long-standing, serious depression that had not responded to treatment. Please do not read this as saying that you should go and treat any mental health conditions you may have with psilocybin you got off a guy at a house party.

homeopathy, is that the sort of person who wants to study psychedelics is often the sort of person who really wants to prove that psychedelics are good. There's a thing in science called the "researcher effect," which is that researchers have an amazing (even if subconscious) tendency to find things they really want to find.

Anyway. The Bayesian model of depression is that it is caused by inappropriately strong priors on some negative belief, perhaps about how you are a bad person or how powerless you are or how bad everything is. (Depression can take many forms.) The metaphor the researchers use is of a "landscape" of beliefs: a landscape of rolling hills and valleys, but also sheer mountains and plunging chasms. "You" are a little car on the landscape. You want to get as low down as possible on the landscape: the lower you are, the "truer" your beliefs are (or to be more precise about it, the more accurately your beliefs match your experience, or the less prediction error you have). You naturally roll downhill, but you can go a bit uphill if you get a "push" with evidence.

Very strong beliefs—"Faces point outward," "The sun will come up tomorrow," that sort of thing—are very deep valleys, with very steep sides. You need a lot of evidence to push you out of them. Weaker beliefs about whether or not your coffee cup has coffee in it can be overcome with less evidence.

The trouble comes when you get stuck in a little local hole that matches your evidence somewhat well, but not as well as a much deeper valley next door. Then you have an "untrue" belief, or, if you prefer, a suboptimal belief that does not predict incoming data as well as an alternative.

That's not such a huge problem if the strength of your belief is commensurate to the evidence. But if your priors are inappropriately strong, then the "valley" will be inappropriately deep, and your little belief-car won't be able to climb the sides, even with lots of good evidence.

That's apparently what might be going on with depression. Your prior probability on some untrue belief, something like "I am a terrible person and everyone hates me," is inappropriately high. Your little car can't get out of the belief valley and into a more accurate valley, in which you are a pretty standard person about whom people have the normal range of opinions.

Evidence that comes in that could prove otherwise—people telling you that you are a nice person and they love you, for instance—gets discounted, because your prior beliefs are so strong that (as you'll remember from the section on multiple hypotheses in chapter 3) the "I am not terrible" explanation is overwhelmed by alternative explanations, such as "This person is lying to make me feel better." It will be essentially impossible to get out of the hole.

"You have an excessive precision-weighting of priors," said Robin Carhart-Harris, a neuroscientist at University of California San Francisco and one of the researchers in the paper mentioned above, when I spoke to him a couple of years ago. "Or, in a more human way, you're too confident in some pathological belief or bias."

Now, psychedelics. They're unusual drugs. They don't particularly make you happy or energetic or anything; they just make you *really interested* in things. They make the world feel unfamiliar. "Have you ever, like, really *looked* at a tree, man?" That sort of thing.

What they do, in this model, is to flatten your priors. You never really *look* at a tree, or your hand, or whatever, because you have very strong prior beliefs about what trees are like, and those beliefs very successfully predict the information that will come from looking at a tree, so your brain basically discounts them. "Familiar thing, accurately predicted, move on."

But if your prior beliefs are made less precise, less confident, then the data coming from your senses will be upweighted. Suddenly the back of your hand is absolutely fascinating. And weird noisy variations in the data—things that your brain normally explains away—get flagged as important things to pay attention to, so you get impressions like the floor is breathing or faces are staring at you from the wallpaper.

This is all stuff you'll remember from the section on schizophrenia earlier in this chapter. Same idea, really. What's important, though, is that *in theory*, if you give a depressed person psilocybin, it will flatten their belief landscape—weaken their inappropriately strong priors on their own terribleness, or whatever the particular belief is. So, combined with therapy that encourages the patient to realize that they are not so awful after all, it allows the little car of belief to leave the valley of depression and move into the "truer" valley in which the patient is not so terrible, where (once the drugs wear off) they will hopefully remain.

(Yes, in theory, you could flatten your priors and move out of a nice true valley into an adjacent, less true one, and cause yourself delusional beliefs. Carhart-Harris told me that was rare, but possible, so it was important to take the drugs under expert supervision.)

As I said, treat this with a certain amount of skepticism. This particular model of depression may or may not be right—I've also seen suggestions that depression can be understood as *under*confidence in neural predictions—and whether psychedelics will end up having any real impact as psychiatric drugs remains to be seen. Even if they do work, there are huge societal and regulatory obstacles—it's hard to get licences to do the research, and prescribing them would be illegal under current laws in the US and the UK. But they're a neat application of the Bayesian brain hypothesis in a real, clinical situation.

GOD HELP US

There's a post by Scott Alexander, himself a psychiatrist, a cult-of-Bayes fanatic, and an all-round very clever man, called "God Help Us, Let's Try to Understand Friston on Free Energy."[34]

Friston, as mentioned, is probably the greatest pioneer of the predictive processing/Bayesian brain model: if you read any scientific papers in the area, you'll keep bumping into references saying "(Friston, 2009)" and "(Friston, 2006)." But his work is also famously difficult to understand. There's even a parody Twitter account, @FarlKriston, dedicated to not understanding him.

Friston takes the Bayesian brain model further. So far we've talked about predictive processing as explaining *how we make sense of the world*—What do these ambiguous nerve signals mean? What's the best way to move my eyes in order to gather information?—that sort of thing. For Friston, though, it explains—or at least describes—a lot more than that. Minimizing prediction error isn't just sense-making. It is, in this model, our fundamental motivation. Hunger, sexual desire, boredom—all our wants and needs—can be described in terms of a struggle to reduce the difference between top-down prediction and bottom-up sense-data, between your prior and your posterior distributions.

And yes, this does *sort of* mean that "being hungry" is the same as "confidently predicting that you are currently eating a sandwich but that prediction being wrong."

More than this: this is, according to Friston, the fundamental driver of *all* life. A bacterium, a mouse, a whale—they're all trying to, in a mathematical sense, reduce the difference between what they predict and what they experience.

Friston talks about "free energy." It's a term from physics—people use it when they're talking about thermodynamics or quantum mechanics. In thermodynamics, it means the amount of energy available to do work, say in a steam engine.

But the same math can be used to describe information theory. In that case free energy is what we've been talking about in this chapter: prediction error. Your brain hates prediction error and wants to minimize it.

It seems incredibly obvious that that's not *all* your brain wants. You don't just care about knowing things. When you jump out of the way of an oncoming bus, it doesn't seem to make sense to say that you are predicting that you're not getting hit by a bus. You just don't want to be hit by a bus. But Friston disagrees.

Imagine a primitive single-celled organism. Its most fundamental goal is to keep the stuff that's inside it different from the stuff that's outside it.

In a sense, that's all that life is. Any system, left to its own devices, tends toward uniformity. A hot drink cools to room temperature, and slightly warms the room as it does so. A cold drink warms up. A balloon slowly deflates until it's the same pressure as the atmosphere. That's *entropy*. Organized systems have low entropy, disorganized ones have high entropy. The universe naturally tends toward entropy. An organized system—like a cold drink in a warm room—becomes disorganized and uniform.

But if a living thing did that, it would die. Being the same as your surroundings is the *same thing* as being dead. If my body reverted to ambient temperatures, if the concentrations of chemicals inside my body were the same as outside my body, then I would no longer exist. That's true of anything living. So any living or self-organizing thing must work to maintain a boundary between itself and the universe. It must maintain the right temperature, the right pressure, the right mix of chemicals on the inside of the boundary. In other words, it must minimize entropy.

A very basic single-celled organism won't make complicated predictions like "Faces tend to point outward." But it needs to maintain chemical concentrations, fluid pressures, temperature, and so on at levels that allow its internal processes to work. It can't directly read them—it behaves like the Kalman filter we mentioned earlier in this chapter. Instead, it relies on indirect evidence. Say, if it's trying to estimate its internal salt concentrations, it might predict the number of sodium ions passing across its cell membrane per second or something. (Not consciously, obviously; in an algorithmic way.)

What's crucial, though, is that the organism can only survive if these predictions are correct. It can't update its model and say, "Ah, looks like I'm severely hyponatremic, better change my predictions of how many sodium ions I expect to pass across my membrane." If it does so, it will rapidly die.

But there are two ways of reducing prediction error. One is changing your prediction, sure. Another is changing the world so it matches your prediction. So the bacterium might metabolize some food, or flail its little flagellum around and get moving until it's somewhere with a higher sodium concentration.

In this model, "desire" and "prediction" are the same thing. The bacterium wants to reduce its prediction error (or "free energy"), for whatever it's predicting. If it happens to be predicting the weather that day, and its predictions are false, then it can update its model, and next time it will make different predictions.

For life-critical predictions, though, its predictions must be fixed. You cannot change your model of what your body temperature is or what your glucose levels are, outside very narrow windows. So the only way of minimizing prediction error is by changing the world, or your position in it, so that your predictions are true.

For Friston, that's what's going on in *all* self-organizing systems. We've been talking about bacteria, but a human has the same thing. We need to maintain homeostasis—a clear distinction between our selves and the universe, and the "self" bit within very specific thermodynamic and chemical limits. What more sophisticated animals, like humans, can do better than bacteria, though, is to manage their surroundings with an eye on the future, to avoid *ending up* in situations where their predictions of "having enough oxygen" or "not being on fire" might stop coming true. In mathematical terms, we want to minimize *expected* prediction error, or *expected* surprise.

"You can talk about *homeostasis* versus *allostasis*," says Friston. Homeostasis, as we've seen, is adjusting your surroundings and your body to maintain a stable internal environment: if your blood sugar drops, your brain orders your pancreas to release more insulin, for instance. Allostasis, he says, is "deliberative planned behavior to avoid having to make sort of homeostatic corrections.

"Let's say I feel hungry," he says. "I'm not hypoglycemic, but say I roll out my plans into the future, for example imagining that I carry on working, then given my model of my own body, I work out that I'll be hypoglycemic in half an hour's time. So I evaluate another plan: I'm going to go and have a nice, sugary, creamy cup of coffee." That other plan reduces the surprise he would expect to feel, because the likeliest future doesn't include his body going into hypoglycemic shock.

To reiterate: according to the free energy model, your brain treats predictions like "I won't get wet if I go outside" and "I will not go into hypoglycemic shock" just the same. It wants to minimize the surprise it receives from those predictions being wrong. But the difference is that if new information comes in that suggests your brain is wrong about getting wet—you see that it's raining, for instance—it has two ways of dealing with it. It can change the world so that its predictions are true, by grabbing an umbrella, or it can change its predictions so that they meet the world, by accepting that you will get wet. It can update its priors.

In the hypoglycemic shock situation, it can't. You have certain, deeply wired priors about the state of the world that will not change. But still, mathematically, it can be treated the same—as prediction error.

These very fundamental priors are wired into us by evolution, Friston says. We don't know exactly which ones they are, although blood sugar levels, body temperature, oxygen levels, and bodily integrity are obvious ones. (Presumably social and sexual desires are hardwired to some extent too, even if they only come online later on.) And as young children, the hardwired priors are the only ones we have—we predict that we will not be hungry, or cold, or injured. "If you're a neonate, you start to learn that when I get these signals, and I cry, Mum appears. They're all these things that have to be learned. And you can learn these preferred states of being that you can work toward, which are constrained by these innate priors underneath them, which keep you alive."

Minimizing free energy means changing your state to avoid prediction error, but it also means trying to find out as much as you can about the world in order to make better predictions: finding the optimal move to make next to gather information, like the Wordle guesses that are intended to rule out letters rather than be a guess in their own right. You can minimize prediction error by generating better models of the world.

There's a process called "motor babbling" that babies do, which is essentially trying out random nerve signals and seeing what happens. Does my leg move? Does my eye twitch? Do I hiccup? "It's a beautiful example of maximizing expected information gain, of learning the nature of the world," says Friston. "What am I in charge of? And what am I not in charge of? Who caused that, did I, did you? They're learning you've got a body and certain things you can control and certain things you can't."

At first, because babies have so little information, their movements are random. They learn as they "babble," and their movements become increasingly sophisticated: they update their priors with each new bit of data. I have a newborn niece, ten weeks old at the time of writing, and you can see this process week by week: her eyes fixing on faces, her hand successfully grasping things. As she finds that she can minimize free energy by doing certain things—moving hands to grasp food; feeding herself with different kinds of food; choosing which brand of pizza to buy—her preferences will become more sophisticated.

"As you get older, and accumulate preferences, you get more skilled in navigating your body through the world," says Friston, "until the point that you can start planning months ahead to meet somebody in a restaurant in a different city."

And this is—according to this model—all exactly the same, mathematically, as what the bacterium is doing when it predicts high sodium ion levels, notes a prediction error, and moves to find more sodium. It's just that our models of the world are deeper and more sophisticated and able to look further ahead.

"The difference between, say, a virus and you and me is how far into the future you can look. We have hierarchically deeper generative models, and what accompanies that is the ability to roll out further into the future."

The way to think of us, says Friston, is as *nearly* perfect scientists. We want to learn about the world, build better and better models of it, looking in the places that will get us the most information, minimizing the difference between the signals we predict we'll receive from the world and the signals we actually do receive. Except that on a few crucial points, we don't want to learn what particular states are like. If we were purely truth-seeking, purely curious, we'd be just as keen to learn what it feels like to hold our hand in a fire, or go without oxygen for two days, as we are to find out what blue cheese tastes like. If we predicted that the best way to gain information about the world would be to stab a fish fork in our eye, then we would do that. But because we have these hardwired prior predictions, and it would lead to huge prediction errors, we won't do that. "We are all crooked scientists," Friston says. We're Bayesian prediction machines, but some of our priors can't be changed, because if they were, we'd die, and dying doesn't help us find things out—so we have to change our environment so that those priors remain correct.

I want to be a bit cautious here. I love the idea of free energy, and I hope I've captured it well—it is, after all, famously complicated. But Friston himself would say, I think, that it's a framework that lets you do math, rather than a scientific theory in its own right. You don't *need* to frame everything in terms of predictions and free energy and information gain—you can just say that we have desires, and those desires often involve not dying. Free energy allows you to simplify the model, use just one term for everything, so it wins Occam points, but that doesn't make it *right*, and some people just find it weird to suggest that hunger is the same as wrongly predicting that you've eaten. But it's an elegant theory.

Now I want to sum all this up, from Bayes in science to the Bayesian brain, and to show you how once you start noticing Bayes' theorem, you see it everywhere.

Bayesian Life

As we said at the beginning: If you think you've found a theory of everything, diagnose yourself with mania and check yourself into a psychiatric hospital. (Mania, of course, can be described in Bayesian terms: some papers suggest that it's to do with pathologically high confidence in your brain's predictions.[1])

Do I need to check myself in? I hope not. But *on the other hand*, I do seem to see Bayes everywhere I look, from the small to the big.

Here's something small. Your email account is Bayesian. If it weren't, your inbox would be even more full of nonsense than it already is. Depending on who you ask, somewhere between 35 and 70 percent of all emails sent in the world are spam: that is, unsolicited ads. (I just checked my Gmail account and I've received ten non-spam and seven spam emails so far this morning, so that's 40 percent, which fits.) Let's say it's 50 percent.

A spam filter takes that as your prior probability and updates with new information. For instance, it might be that 20 percent of spam emails contain the phrase "penis extension," whereas only 5 percent of non-spam emails do.

So if your filter were to see a million emails, it would expect to see five hundred thousand spam and five hundred thousand not-spam emails. Of the spam ones, about one hundred thousand would contain the phrase "penis extension," while of the non-spam ones, about twenty-five thousand would. Your spam filter would, therefore, judge that a given email with the phrase "penis extension" in it has an 80 percent chance of being spam. If it contains words like "act now," "porn," or "low-interest loan," your filter would further update. If an email reaches a certain threshold for spam-probability, it will get bumped into your spam folder. This is explicitly how spam filters work: Google "naive Bayes spam filtering."

Now here's something big: evolution. The astronomer Fred Hoyle once said that the chances of evolution successfully producing life were similar to the chances of a whirlwind passing through a junkyard and creating a Boeing 747.[2] But he misunderstood evolution, which is not random. He was right that the number of ways in which the component parts of a 747 could be arranged are unfathomably huge; if you put them together at random, the chances of making something that could fly are tiny. Similarly, if you took apart the body of, say, a fruit bat, right down to its cells, and then put them back together at random, the chances of making something that flew (and fed itself, and reproduced) would be infinitesimal.

In reality, though, evolution does not put things together at random. It searches through the space of possible arrangements, through the non-random process of natural selection. If you have a simple, self-replicating thing that makes copies of itself with occasional small, random mistakes, then copies that are better at replicating will tend to make more copies; copies that are worse will tend to be eliminated.

You can (of course) see this as a Bayesian process. Remember the beeping box that told you (imperfectly) whether you'd got a winning lottery ticket. That allowed you to search the space of lottery numbers. At first, as far as you're concerned, all 131,115,985 are equally likely to be the winning one; after you've run the box over them all, you narrow your search space to a quarter the size. It's an optimization process, moving through the vast space of possibilities to reach the target you actually want.

Evolution works in just the same way, albeit much less efficiently. There's an equation, Price's equation, which says that the frequency of some characteristic in a population will change according to how much that characteristic is related to "relative fitness," i.e., how well the organism reproduces. To take a simple example: if gazelles that can run faster are more likely to survive because they are less likely to be eaten by lions, then on average, you'll expect to see more fast-moving gazelles surviving through to the next generation.

You can see the genome of an organism as a "prediction" about the world. If your genes build fast-running legs, they're predicting you will be born into an environment full of fast-running predators (or prey). If your genes build a short, tough beak for cracking nuts, they're predicting you will be born into an environment that contains a lot of nuts. If your genes build long, piercing canines for gripping wildebeests' jugulars, it's a prediction that your environment will contain wildebeests. It's also a prediction that the rest of your genes will form a body that is useful for those attributes: piercing canines would be no use to an earthworm or a yew tree.

The frequency of a gene in a population encodes a prior probability. The new data, the likelihood, come when the organisms built by the genome containing that gene either survive and reproduce or don't. If many copies of the gene make it into the next generation, that's evidence that the gene is suited to its environment. If few do, that's evidence that it's not. Weak evidence, either way—a gene can be deleterious to survival but survive through lucky association with other genes, or it can be very useful but through misfortune be born into a creature that gets hit by an avalanche—but evidence, nonetheless. Evolution is slow, blind, and inefficient—it might take hundreds of generations to solve a problem a human designer could solve in an hour—but it's an approximation of a Bayesian process: it works to minimize prediction error.

But then, everything decision-related is Bayesian. It simply describes the optimal way to integrate new information with your prior best guesses. When you look at it like that, so many things seem to make more sense. Confirmation bias, for instance: we're told that people are more likely to trust evidence that supports what they already believe. It sounds bad, and sometimes it can be. But most of the time, it's just good Bayesian reasoning. If my friend tells me they saw a fox in North

London, I'll probably believe them, because foxes are pretty common in North London. If my friend tells me they saw a Cape buffalo, I'll assume they're joking, or unwell, unless they provide me with some pretty solid evidence. The only difference is that my prior probability on the presence of Cape buffaloes is low.

Now, as it happens, I think that prior probability estimate is pretty good, and most people would probably agree. But when we talk about confirmation bias, we're usually talking about disagreements where people disagree quite strongly. If you have a strong prior that vaccines cause autism, then you will be more skeptical of evidence that says they don't. Most people reading this would probably think that prior is inappropriate, but if you have it, it might take a lot of very good evidence to move, and—if it's strong enough—it might be all but impossible to shift it, because alternative hypotheses such as "mainstream science is lying to us" start out with higher probabilities.

It also explains why we might trust some people more than others. You see studies every so often saying that people will judge the same speech differently if they're told it was given by a Republican than if told it was given by a Democrat, and this is seen as a demonstration of our fundamental irrationality. But again it's perfectly rational, if you think that people have different priors on the trustworthiness of different political parties, and therefore of people associated with those parties. For a Republican, Pete Buttigieg or Joe Biden saying something might be a low-precision update, a wide, flat likelihood curve on the graph, and would shift the Republican's prior beliefs only somewhat. It could even be *anti*correlated: you might be *less* likely to believe some controversial statement if someone you deeply distrust claims it.

Or: a study came out toward the end of 2022[3] that found that peer reviewers in science were more likely to accept papers for publication if they saw that they were written by Nobel-winning authors than if they were written by novices. That might not be ideal—in theory, at least, science shouldn't rest on reputations—but it's rational: if you have two scientific papers in front of you, and you know nothing else about them other than that one was written by Albert Einstein and the other by Cletus B. Nobody, you'll have a higher prior probability that the first one will be good. Once you've read the papers, if they both seem pretty good to you, then you'll update toward accepting them. But unless you have *total* confidence in your ability to judge the paper entirely on its merits, then the new data won't completely wash out your prior probability, and you'll still judge the Einstein paper as more likely to be good.

The statistician George Box, he who sang "There's No Theorem Like Bayes' Theorem" at the first Valencia conference, had a saying: "All models are wrong, but some are useful."[4] He was thinking of statistical models, of the economy, or of climate change, or whatever. You can model the behavior of a gas with the ideal gas law, he said, and it will be wrong—it won't perfectly match what happens—but it might be close enough to be useful.

But the point is wider. We all have models of the world in our heads. The models contain quotidian things like doors and spouses and coffee shops, and more abstruse things like planetary orbits and international trade and viral vectors of transmission.

Models make predictions. Mine predicts that the door is behind me and will open when I twist the handle; that my wife would rather watch *Moonlight* than *Dune* once the kids have gone to bed; that the

XBB.1.5 Omicron variant, which, at the time of writing, is dominant in the United States, will also become dominant in the UK, but will not cause a major wave of deaths and hospitalizations in our highly vaccinated population.

All those models will be imperfect. I won't have exactly the right weight of the door in my mind, and my model of my wife's tastes and preferences is revealed to be indifferent at best every Christmas. My understanding of how viruses spread and of the human immune system is thin. But the extent to which those models are any good is the extent to which they predict the world. And I update them each time new information comes in. If XBB.1.5 *does* cause a wave of severe disease, then I'll have to reassess my model.

So it's all predictions, and the interesting thing is prediction error. A confident, precise prior that is contradicted by precise information coming from the world should result in a radically changed posterior probability. And the degree to which you should change your beliefs is dictated by Bayes' theorem.

What I hope I've shown in this book is that this is true all the way up and down. It's true of conscious, explicit predictions like the one about the new variant or whether a soccer team will win, but it's also true of our informal predictions of other people's behavior or how a ball will bounce.

Fascinatingly, it's also true much more deeply. Our perception of the world is one of constant predictions, tested against the evidence from our senses. We hypothesize that a small point of light is from a small, nearby object, or that a certain mid-gray color on our retina is caused by a dark object under bright light. We test those predictions by gaining new information, by moving our heads or inspecting the object. And even at the very lowest level, the human brain seems to work by predicting the number and pattern of nerve firings it expects, rewarding itself with dopamine when those predictions are close to reality, and punishing itself with an absence of expected dopamine when they're not.

The exact details of the hypothesis may change. Perhaps the active inference/predictive processing model is wrong in some important way. But the brain clearly does work by making predictions and updating them. And what's more, this hypothesis makes so many things make sense. Optical and auditory illusions, hallucinations, dreaming, mental illness.

There's a thing called a "flow state," when you're doing some activity, playing an instrument, playing sports or a video game, painting, whatever, and it just seems to *work*: that's when your predictions are high-precision and they're coming true every time. When on a dark street you briefly mistake a mailbox for a person, that's your brain forming a hypothesis from noisy data. When you find your eyes suddenly drawn to someone with a facial disfigurement, you're not just being rude: your brain has strong priors about what faces look like, and when those priors are confounded and it receives prediction errors, it seeks more information.

It explains why, as we get older, we get more set in our ways. When we're young, we have very little data about the world, so our priors are weak and new information can shift them easily. We can learn quickly, because we don't have a very precise model of the world that makes good predictions. As we get older, though, we gain more information, we get a richer, more precise model of the world, and new information must logically shift our priors less. So older people (to quote Friston) "are wise, but inflexible." You can predict the world much more accurately when you're older, as long as the world doesn't change. But if the world does change, you need much more information in order to shift your preexisting beliefs. Hence the tropes that dads end up getting their kids to help them set the video recorder.

Even consciousness itself makes more sense in a Bayesian framework. We can think of our experience of the world as our predictions of the world, our Bayesian prior. It doesn't solve the hard problem of consciousness, but it seems to give us an interesting place to look.

This prediction-testing model also fits in perhaps the highest-level thinking that humans do: science. Science is explicitly about making predictions—hypotheses—and testing them; Helmholtz and Gregory used science as a model for human perception. The problem is that, in science, we like to think that there is an objective truth out there, and the Bayesian model of perception is explicitly subjective. A probability estimate isn't some fact about the world, but my best guess of the world, given the information I have.

But if we want to ask questions like *How likely is it that my hypothesis is true, given this new data?* we have to use prior probabilities—we have to be Bayesian—and the only way of having prior probabilities is to use subjective estimates. That doesn't mean they can be pulled out of thin air—there are more and less reasonable priors to hold, and we can check whether ours are reasonable by crowdsourcing them and checking whether our discoveries would stand up if we'd started with somewhat different priors. But they're still our imperfect guesses about the real underlying facts of the world.

That doesn't mean that science can't know anything, or that it's all perfectly postmodern. It just means that, once again, we're building models of the world, and trying our best to check them against the real world: making predictions and updating them with new information, trying to minimize prediction error. We have a mental map, and the map is not the territory, but the territory exists, and if the map is wrong it will send us to the wrong places.

In fact, a Bayesian model like this seems a neat way of thinking about science. Philosophers of science get bogged down with epistemology. We can't know anything for certain; we might be being deceived by an evil demon, we might be a brain in a jar. We might see a million white swans, but we can't be sure we'll never see a black swan, so can we really ever say, "All swans are white"? It's easy to end up in weird places if you go too far, like Paul Feyerabend or Robert Anton-Wilson saying that all knowledge is impossible. Or Popper saying that there's no such thing as confirming a theory, only disproving it. But obviously knowledge *is* possible, or at least we can make reliable predictions of the world. I very confidently predict that an airplane will take off and land successfully thanks to the laws of aerodynamics.

But it's very straightforward with Bayes. I have a hypothesis about what percentage of swans are white, and I test it against the evidence. Maybe I start by estimating that 50 percent of swans are white, but as I see more and more white swans, I push my probability distribution until I have quite a lot of my probability mass on "All swans are white."

But I never get to total certainty, just as Thomas Bayes gets more confidence in the whereabouts of the white ball as more and more red ones are thrown, without ever being sure.

Then, if I see counterexamples, it immediately becomes far less likely that all swans are white (although not impossible: maybe I hallucinated) and makes me move my probability distribution along. We're not forced into strange postmodern positions where all models of the world are equally valid; we can be empirical, we can say that the heliocentric model of the solar system predicts the world more accurately than the geocentric model or that the most-swans-are-white hypothesis is closer to the truth than the all-swans hypothesis. But we're comfortable with uncertainty, with never saying we have the absolute final answer.

Whether that means we ought to do the *statistics* of science in a Bayesian way is a separate question. I don't think it would solve every problem, and it's probably more appropriate in some places than others—as Daniël Lakens said, when you get five-sigma results from the Large Hadron Collider, your priors don't matter all that much. But it does avoid some of the problems of frequentist science, and once again it lets us think in terms of how confident we are in some hypothesis, rather than simply accepting or rejecting it. And it has an aesthetically pleasing neatness to it.

But perhaps this is all a bit theoretical. Your brain does what it does, and science can use whatever statistics it feels most comfortable with. But I think there are lessons we can all draw from Bayesianism: practical benefits we can gain from taking some of the ideas and employing them in our lives. I'm not saying run Bayes' rule over every belief, but keep a couple of things in mind.

First, you don't need to think so much in terms of *right* and *wrong*, *true* or *false*. You can think in terms of how confident you are in a belief, and adjust it up and down, rather than rejecting or accepting it at some arbitrary threshold.

Most of us either believe things or we don't. And that means when evidence comes in that contradicts some belief, we have to either reject the evidence or change the belief. But if we think in terms of percentage probabilities, we can incorporate new evidence, allow it to move our probability distribution up or down.

And the opposite is true. When we read some new scientific study, say, claiming that red wine causes cancer, then we don't have to just believe or disbelieve it. We can think, "What's my prior probability? How likely do I think this is?" Maybe you don't need to use explicit percentage probabilities, but you can use your own knowledge of the world, and allow the new information to adjust it, rather than being blown here and there with every new piece of information.

One thing I find Bayesianism particularly useful for is short-circuiting arguments about whether someone is lucky or skillful. Warren Buffett has made a lot of money investing in the stock market. Does that make him a good investor? Or has he just got lucky? Or Bill Ackman. He's made something like $4.5 billion. The same in sports: Is a golfer or a tennis player lucky to have won? If I intend to sink a pool shot, and then I do, can I claim it was skill?

If you're allowed to think in Bayesian terms, you avoid this unnecessary binary. With no information at all, my prior is that any given investor will be somewhat worse than the stock market at picking stocks. But each time that investor successfully picks one—or each time a golfer makes par or wins a tournament—I upgrade my confidence in their skill and reduce the probability that they just got lucky. Perhaps Warren Buffett has been consistently lucky for fifty years, but I suspect it's unlikely.

The idea that beliefs are *predictions* is key, as well. Remembering that allows you to sidestep an awful lot of the worst arguments in the world. For instance: A lot of ink gets spilled on the question "Is cancel culture real?" But most of the people arguing it agree on the actual facts at hand—some people have lost their jobs over things they said on the internet—and just disagree on whether those facts deserve to be called "cancel culture." If you agreed that "cancel culture" was real, or wasn't, would that change any predictions that you make about the future? What prediction error could you experience that would make you up- or downgrade your confidence in that belief? If there isn't one, then maybe you're just arguing about the definition of a word, rather than about any real claim about things out there in the universe, and you can forget about it and start talking about more concrete things instead.

And you'll start to notice that an awful lot of arguments—in the real world between friends, and in the media, and online—are about whether we should use some word to describe some set of phenomena: Is it woke? Is it racist? Is it eugenics? But very often, nothing hinges on the outcome of the argument beyond the label you get to attach to something. Perhaps that's useful if you want to win some argument or gather support for some political action, like banning something, but it doesn't change the predictions you make about the world.

As we said at the beginning: you can predict the future. You do it every single second. You're doing it at a micro-level, and have to, if you are to successfully navigate the world and not trip every time you try to walk. You're doing it at a very high level when you book a holiday for next year and predict that Lanzarote will still exist and that Jet2's Airbus will fly you there. And you're doing it in all sorts of intermediate ways, when you go to a store predicting that they'll have craft IPA or chocolate cookies, or when you avoid mentioning your friend's recent divorce because you predict it will upset them. There's nothing mystical about it: that's just how we work. Humans are prediction machines. And Thomas Bayes showed us the math of how we do it.

Acknowledgments

I couldn't have written this book without the help of a lot of people who, to put it bluntly, understand Bayes' theorem, or some application or aspect of it, a lot better than I do. In alphabetical order, they are: David Bellhouse, Sophie Carr, Corey Chivers (no relation), David Chivers (some relation), Nikitas Chrysaitis, Aubrey Clayton, Paul Crowley, Pete Etchells, Alexandra Freeman, Chris Frith, Andy Grieve, Jonny Kitson, Daniël Lakens, Jens Koed Madsen, David Manheim, Marcus Munafo, Peggy Series, Anil Seth, Murray Shanahan, Michael Story, Helen Toner, Julia Wise, and William Woof.

Breaking the alphabetical order thing, for the second book in a row I'd like to give Kevin McConway, emeritus professor of statistics at the Open University, a particular mention, because he once again read through the book and gently pointed out several places in which I had entirely misunderstood the point. Without his intervention there would be many more errors in this book than there are, which is probably a lot, despite his best efforts. I sent him a bottle of Ardbeg as thanks, but I felt some public acknowledgment would be in order too.

Jenny Lord, Lucinda McNeile, and the rest of the team at Weidenfeld & Nicolson, plus Will Francis at Janklow & Nesbit, made the book actually happen, which is probably a good thing, on balance.

My talented sister, Sarah Chivers, once again drew the illustrations, and once again they look great.

Claire Trumble and Marcus McGillycuddy kindly gave their permission for me to dedicate this book to the memory of Luis.

And, of course, I thank my wife, Emma, and children, Billy and Ada, for being splendid.

Notes

INTRODUCTION: A THEORY OF NOT QUITE EVERYTHING

1. Scott Alexander, "Book Review: Surfing Uncertainty," *Slate Star Codex*, September 5, 2017, https://slatestarcodex.com/2017/09/05/book-review-surfing-uncertainty.
2. Nick Collins, "Stephen Hawking: Ten Pearls of Wisdom," *Telegraph*, September 3, 2010, https://www.telegraph.co.uk/news/science/science-news/7978898/Stephen-Hawking-ten-pearls-of-wisdom.html.
3. H. P. Beck-Bornholdt and H. H. Dubben, "Is the Pope an Alien?," *Nature* 381 (1996): 730, https://doi.org/10.1038/381730d0.
4. S. J. Evans et al., "Prevalence of Adult Huntington's Disease in the UK Based on Diagnoses Recorded in General Practice Records," *Journal of Neurology, Neurosurgery & Psychiatry* 84 (2013):1156–60.
5. M. Alexander Otto, "FDA Grants Emergency Authorization for First Rapid Antibody Test for COVID-19," Medscape, April 4, 2020, https://www.medscape.com/viewarticle/928150.
6. John Redwood, (@johnredwood), "The government advisers today need to tell us how they are going to stop false test results distorting the figures," Twitter, September 21, 2020, https://twitter.com/johnredwood/status/1307921384883073024.

7. "What Should I Advise about Screening for Prostate Cancer?," NICE, last updated February 2022, https://cks.nice.org.uk/topics/prostate-cancer/di agnosis/screening-for-prostate-cancer.

8. P. Rawla, "Epidemiology of Prostate Cancer," *World Journal of Oncology* 10, no. 2 (April 2019): 63–89, doi: 10.14740/wjon1191.

9. H. D. Nelson et al., "Harms of Breast Cancer Screening: Systematic Review to Update the 2009 U.S. Preventive Services Task Force Recommendation," *Annals of Internal Medicine* 164, no. 4 (February 16, 2016): 256–67, doi: 10.7326/M15-0970.

10. "Breast Screening," NICE, last updated May 2022, https://cks.nice .org.uk/topics/breast-screening/.

11. S. Taylor-Phillips et al., "Accuracy of Non-Invasive Prenatal Testing Using Cell-Free DNA for Detection of Down, Edwards and Patau Syndromes: A Systematic Review and Meta-Analysis," *BMJ Open* 6 (2016): e010002, doi: 10.1136/bmjopen-2015-010002.

12. C. Jowett, "Lies, Damned Lies, and DNA Statistics: DNA Match Testing, Bayes' Theorem, and the Criminal Courts," *Medicine, Science and the Law* 41, no. 3 (2001): 194–205, doi: 10.1177/002580240104100302.

13. Steven Strogatz, "Chances Are," *New York Times*, April 25, 2010, https:// archive.nytimes.com/opinionator.blogs.nytimes.com/2010/04/25 /chances-are/.

14. Gerd Gigerenzer, *Reckoning with Risk: Learning to Live with Uncertainty* (London: Penguin, 2003), 141.

CHAPTER ONE: FROM *THE BOOK OF COMMON PRAYER* TO THE FULL MONTE CARLO

1. T. Bayes and R. Price, "An Essay towards Solving a Problem in the Doctrine of Chances. By the Late Rev. Mr. Bayes, F. R. S. Communicated by Mr. Price, in a Letter to John Canton, A. M. F. R. S.," *Philosophical Transactions* 53, no. 1763 (1683–1775): 370–418.

2. D. R. Bellhouse, "The Reverend Thomas Bayes, FRS: A Biography to Celebrate the Tercentenary of His Birth," *Statistical Science* 19, no. 1 (2004): 3–43, https://doi.org/10.1214/088342304000000189.

3. Much of this history is taken from Bellhouse's short biography of Bayes,

and from conversations with Bellhouse himself. I am grateful to him for this scholarship and commend the work, which is available for free online.

4. J. Landers, *Death and the Metropolis: Studies in the Demographic History of London, 1670–1830* (Cambridge, UK: Cambridge University Press, 1993), 136.

5. Stephen Stigler, "Richard Price, the First Bayesian," *Statistical Science* 33, no. 1 (February 2018): 117–25.

6. T. Birch, *An Account of the Life of John Ward, LL.D., Professor of Rhetoric in Gresham College; F.R.S. and F.S.A.* (London: P. Vaillant, 1766), quoted in Bellhouse, "The Reverend Thomas Bayes."

7. G. A. Barnard and T. Bayes, "Studies in the History of Probability and Statistics: IX. Thomas Bayes's Essay towards Solving a Problem in the Doctrine of Chances," *Biometrika* 45, no. 3/4 (1958): 293–315, https://doi.org/10.2307/2333180.

8. *Letters to Thomas Bayes and John Skinner Smith in Ward's Latin Correspondence*, British Library Manuscript, Additional Manuscript 6224, 116. Cited in Bellhouse, "The Reverend Thomas Bayes."

9. Alexander Gordon, "Peirce, James," *Dictionary of National Biography, 1885–1900*, https://en.wikisource.org/wiki/Dictionary_of_National _Biography,_1885-1900/Peirce,_James.

10. T. Bayes, *Divine benevolence: Or, an attempt to prove that the principal end of the divine providence and government is the happiness of his creatures: being an answer to a Pamphlet, entitled, Divine rectitude; or, An Inquiry concerning the Moral Perfections of the Deity. With a refutation of the notions therein advanced concerning beauty and order, the Reason of Punishment, and the Necessity of a State of Trial antecedent to perfect Happiness* (London: Printed for John Noon, at the White-Hart in Cheapside, near Mercers-Chapel, 1731).

11. David Hume, *Dialogues Concerning Natural Religion*, 187, via the Gutenberg Project, https://www.gutenberg.org/files/4583/4583-h/4583-h.htm.

12. John Balguy, *Divine Rectitude: Or, a Brief Inquiry Concerning the Moral Perfections of the Deity; Particularly in Respect of Creation and Providence* (London: Printed for John Pemberton, at the Buck, over-against St. Dunstan's Church, Fleetstreet, 1730).

13. Bellhouse, "The Reverend Thomas Bayes," 10.

14. James Foster, An Essay on Fundamentals in Religion (1720), taken from

Unitarian Tracts in Nine Volumes (London: British and Foreign Unitarian Association, 1836).

15. D. Coomer, *English Dissent under the Early Hanoverians* (London: Epworth Press, 1946), quoted in Bellhouse, "The Reverend Thomas Bayes."

16. Bellhouse, "The Reverend Thomas Bayes," 12.

17. Ibid., 13.

18. E. Montague, *The Letters of Mrs. Elizabeth Montagu, with Some of the Letters of her Correspondents*, vols. 1–4 (1974; London: T. Cadell and W. Davies, London, 1809–13), quoted in Bellhouse, "The Reverend Thomas Bayes."

19. Thomas Bayes, *An introduction to the doctrine of fluxions, and defence of the mathematicians against the objections of the author of the Analyst, so far as they are designed to affect their general Methods of Reasoning*, 1736.

20. David Bellhouse, personal conversation, 2022.

21. Centre for Kentish Studies, Stanhope of Chevening Manuscripts: U1590 /C21—Papers addressed to or collected by Lord Stanhope; U1590/C14/2.

22. J. Lagrange, *Oeuvres de Lagrange, Publiées par les Soins de M. J.-A*, Serret 3 (1869): 441–76; Serret 5 (1870): 663–84 (Paris: Gauthier-Villars), cited in Bellhouse, "The Reverend Thomas Bayes."

23. P. Gorroochurn, "The Chevalier de Méré Problem I: The Problem of Dice (1654)," in *Classic Problems of Probability*, ed., P. Gorroochurn (Hoboken, NJ: John Wiley & Sons, 2012), 14, https://doi.org/10.1002 /9781118314340.ch3.

24. The letters between Pascal and Fermat are available in full at https://www .york.ac.uk/depts/maths/histstat/pascal.pdf.

25. The text of Pacioli's work, and that of Cardano and Tartaglia, is quoted in Jim Sauerberg, *The Problem of the Points: Core Texts in Probability* (Moraga, CA: Saint Mary's College, 2012), http://math.stmarys-ca.edu/wp-content /uploads/2015/08/prob-talk.pdf.

26. Prakash Gorroochurn, "Some Laws and Problems of Classical Probability and How Cardano Anticipated Them," *Chance* 25, no. 4 (2012): 13–20, doi: 10.1080/09332480.2012.752279.

27. Example taken from Aubrey Clayton, *Bernoulli's Fallacy: Statistical Illogic and the Crisis of Modern Science* (New York: Columbia University Press, 2021), 7.

28. Jakob Bernoulli, *Ars conjectandi, opus posthumum. Accedit Tractatus de seriebus infinitis, et epistola gallicé scripta de ludo pilae reticularis* (Basel, Switzerland: Thurneysen Brothers, 1713), trans. Oscar Sheynin (Berlin, 2005) pt. 4, 19, http://www.sheynin.de/download/bernoulli.pdf.

29. Quoted in G. Gigerenzer et al., *The Empire of Chance: How Probability Changed Science and Everyday Life* (Cambridge, UK: Cambridge University Press, 1989).

30. J. Piaget and B. Inhelder, *The Origin of the Idea of Chance in Children*, trans. L. Leake Jr., P. Burrel, and H. D. Fishbein (1951; New York: Norton, 1975).

31. S. Raper, "Turning Points: Bernoulli's Golden Theorem," *Significance* 15 (2018): 26–29, https://doi.org/10.1111/j.1740-9713.2018.01171.x.

32. Clayton, *Bernoulli's Fallacy*, 74.

33. Stephen Stigler, *The History of Statistics: The Measurement of Uncertainty before 1900* (Cambridge, MA: Harvard University Press, 1986), 117.

34. Plato, *The Republic*, book 7, trans. Benjamin Jowett, 198, http://www.filepedia.org/files/Plato%20-%20The%20Republic.pdf.

35. Bernoulli, *Ars Conjectandi*, book 4, ch. 1.

36. Stigler, *The History of Statistics*, 107.

37. Abraham de Moivre, *The Doctrine of Chances: Or, A Method of Calculating the Probability of Events in Play* (London: W. Pearson, 1718).

38. Taken from Stigler, *The History of Statistics*, 124.

39. Biography by Niccolò Guicciardini, in *Oxford Dictionary of National Biography, 2001–2004* (Oxford, UK: Oxford University Press, 2004), cited at https://mathshistory.st-andrews.ac.uk/Biographies/Simpson/.

40. Ibid.

41. Ibid.

42. T. Simpson, "A Letter to the Right Honorable George Earl of Macclesfield, President of the Royal Society, on the Advantage of Taking the Mean of a Number of Observations in Practical Astronomy," *Philosophical Transactions of the Royal Society of London* 49 (1755): 82–93.

43. Stigler, *The History of Statistics*, 138.

44. David Bellhouse, personal conversation, 2022.

45. Letter from Thomas Bayes to John Canton, undated but likely from 1755, cited in Bellhouse, "The Reverend Thomas Baines," 20.

46. Thomas Simpson, *Miscellaneous Tracts on Some Curious, and Very Interesting Subjects in Mechanics, Physical-Astronomy, and Speculative Mathematics* (London: John Nourse, 1757), 64.

47. David Spiegelhalter, *The Art of Statistics: Learning from Data* (New York: Pelican, 2019), 306.

48. Stigler, *The History of Statistics*, 180.

49. Spiegelhalter, *The Art of Statistics*, 324.

50. Bayes and Price, "An Essay towards Solving a Problem."

51. Example taken from Spiegelhalter, *The Art of Statistics*.

52. Ibid., 325.

53. Stigler, *The History of Statistics*, 179.

54. From the will of Thomas Bayes, cited in Barnard and Bayes, "Studies in the History of Probability."

55. Letter to Thomas Jefferson from Richard Price, July 2, 1785, https://founders.archives.gov/documents/Jefferson/01-08-02-0197; letter from Thomas Jefferson to Richard Price, August 7, 1785, https://founders.archives.gov/documents/Jefferson/01-08-02-0280; and letter from Thomas Jefferson to Richard Price, January 8, 1789, https://founders.archives.gov/documents/Jefferson/01-14-02-0196; all in the Library of Congress.

56. Letter from Benjamin Franklin to Richard Price, October 9, 1780, https://founders.archives.gov/documents/Franklin/01-33-02-0330; and letter from Benjamin Franklin to Richard Price, October 9, 1780, https://founders.archives.gov/documents/Franklin/01-41-02-0002; both stored in the Library of Congress.

57. Thomas Fowler and Richard Price, *Dictionary of National Biography*, vol. 46 (1896), 335.

58. David Bellhouse, personal conversation.

59. David Bellhouse, "On Some Recently Discovered Manuscripts of Thomas Bayes," *Historia Mathematica* 29 (2002): 383–94.

60. Stephen M. Stigler, "Richard Price, the First Bayesian," *Statistical Science* 33, no. 1 (February 2018): 117–25.

61. Bayes and Price, "An Essay towards Solving a Problem."

62. David Hume, "Of Miracles," in *Philosophical Essays Concerning Human Understanding* (London: Millar, 1748), 83.

63. R. Price, *Four Dissertations* (London: Millar and Cadell, 2nd ed. 1768, 3rd ed. 1772, 4th ed. 1777).

64. R. Price, *Four Dissertations*, cited in Stigler, "Richard Price, the First Bayesian."

65. Hume to Price, March 18, 1767, in D. O. Thomas and B. Peach, *The Correspondence of Richard Price, Vol. I: July 1748–March 1778* (Durham, NC: Duke University Press, 1983), 45–47, cited in Stigler, "Richard Price, the First Bayesian."

66. David Bellhouse and Marcio Diniz, "Bayes and Price: When Did It Start?," *Significance* 17, no. 6 (December 2020): 6–7, https://doi.org/10.1111/1740-9713.01460.

67. Bernoulli, *Ars Conjectandi*, 19, cited in Clayton, *Bernoulli's Fallacy*.

68. Laplace (1786), 317–18, cited in Stigler, *The History of Statistics*.

69. Clayton, *Bernoulli's Fallacy*, 120.

70. Stigler, *The History of Statistics*, 242.

71. Francis Edgeworth, "The Philosophy of Chance," *Mind* 31 (1922): 257–83.

72. Louis-Adolphe Bertillon, *Dictionnaire Encyclopédique des Sciences Medicales*, 2nd series, 10 (Paris: Masson & Asselin, 1876), 296–324, cited in Stigler, *The History of Statistics*.

73. George Boole, *An Investigation of the Laws of Thought on Which Are Founded the Mathematical Theory of Logic and Probabilities* (London: Walton and Maberly, 1854), 370.

74. Stigler, *The History of Statistics*, 362.

75. Francis Galton, *Natural Inheritance* (London: Macmillan, 1894), 64.

76. Francis Galton, "Regression towards Mediocrity in Hereditary Stature," *Journal of the Anthropological Institute of Great Britain and Ireland* 15 (1886): 246–63, https://doi.org/10.2307/2841583.

77. R. Plomin and I. J. Deary, "Genetics and Intelligence Differences: Five Special Findings," *Molecular Psychiatry* 20, no. 1 (February 2015): 98–108, doi: 10.1038/mp.2014.105.

78. Francis Galton, "Hereditary Talent and Character," *Macmillan's Magazine* 12 (1865): 157–66, 318–27.

79. Clayton, *Bernoulli's Fallacy*, 133.

80. Francis Galton, letter to the editor of the London *Times*, June 5, 1873.

81. Bradley Efron, "R. A. Fisher in the 21st Century," *Statistical Science* 13, no. 2 (1998): 95–114.

82. H. E. Soper et al., "On the Distribution of the Correlation Coefficient in

Small Samples. Appendix II to the Papers of 'Student' and R. A. Fisher. A cooperative study," *Biometrika*, 11 (1917): 328–413, https://doi.org /10.1093/biomet/11.4.328.

83. Ronald A. Fisher, "Some Hopes of a Eugenicist," *Eugenics Review* 5, no. 4 (1914): 309.

84. Karl Pearson and Margaret Moul, "The Problem of Alien Immigration into Great Britain, Illustrated by an Examination of Russian and Polish Jewish Children: Part II," *Annals of Eugenics* 2, no. 1–2 (1927): 125.

85. K. Pearson and M. Moul, (1925), "The Problem of Alien Immigration into Great Britain, Illustrated by an Examination of Russian and Polish Jewish Children," *Annals of Eugenics* 1 (1925): 5–54, https://doi.org /10.1111/j.1469-1809.1925.tb02037.x

86. John Stuart Mill, *A System of Logic, Ratiocinative and Inductive*, vol. 2 (1843), 71.

87. Joseph Bertrand, *Calcul des probabilités* (Gauthier-Villars, 1889), 5–6.

88. John Venn, *The Logic of Chance* (London: Macmillan, 1876), 22.

89. Ronald A. Fisher, "On the Mathematical Foundations of Theoretical Statistics," *Philosophical Transactions of the Royal Society of London. Series A, Containing Papers of a Mathematical or Physical Character* 222 (1922): 312.

90. Ronald Aylmer Fisher, "Uncertain Inference," *Proceedings of the American Academy of Arts and Sciences* 71, no. 4 (1936): 245–58, https://doi.org /10.2307/20023225.

91. This section draws on Sandy Zabell, "R. A. Fisher on the History of Inverse Probability," *Statistical Science* 4, no. 3 (1989): 247–56.

92. George Boole, "On the Theory of Probabilities," *Philosophical Transactions of the Royal Society of London* 152 (1862): 225–52.

93. R. A. Fisher, "Inverse Probability," *Mathematical Proceedings of the Cambridge Philosophical Society* 26 (1930): 528–35, doi: 10.1017/S0305004 100016297.

94. R. A. Fisher, "On the Mathematical Foundations of Theoretical Statistics," *Philosophical Transactions of the Royal Society of London*, Ser. A 222 (1921): 309–68.

95. R. A. Fisher, *Statistical Methods for Research Workers* (Edinburgh: Oliver and Boyd, 1925), 10.

96. R. A. Fisher, "The Arrangement of Field Experiments," *Journal of the Ministry of Agriculture* 33 (1926): 504, https://doi.org/10.23637 /rothamsted.8v61q.

97. Much of what follows is drawn from Sharon Bertsch McGrayne, *The Theory That Would Not Die: How Bayes' Rule Cracked the Enigma Code, Hunted Down Russian Submarines, and Emerged Triumphant from Two Centuries of Controversy* (New Haven, CT: Yale University Press, 2011).

98. H. Jeffreys, *Scientific Inference* (Cambridge, UK: Cambridge University Press, 1931), reprinted with Addenda 1937, 2nd modified edition 1957, 1973.

99. H. Jeffreys, "The Rigidity of the Earth's Central Core," *Geophysical Journal International* 1 (1926): 371–83, https://doi.org/10.1111/j.1365-246X .1926.tb05385.x.

100. David Howie, *Interpreting Probability: Controversies and Developments in the Early Twentieth Century* (Cambridge, UK: Cambridge University Press, 2002) 126.

101. Cited in Ibid.

102. D. V. Lindley, "Sir Harold Jeffreys," *Chance* 4, no. 2 (1991): 10–21, doi: 10.1080/09332480.1991.11882423.

103. Ibid.

104. F. P. Ramsey, "Truth and Probability," *Studies in Subjective Probability*, ed., H. E. Kyburg, H. E. Smokler, and E. Robert (Huntington, NY: Krieger, 1926), 183.

105. Ibid., 65.

106. Cheryl Misak, *Frank Ramsey: A Sheer Excess of Powers* (Oxford, UK: Oxford University Press, 2020), 271.

107. The following examples are taken from McGrayne, *The Theory That Would Not Die.*

108. José M. Bernardo, "The Valencia Story: Some Details on the Origin and Development of the Valencia International Meetings on Bayesian Statistics," *ISBA Newsletter*, December 1999, https://www.uv.es/bernardo/Va lenciaStory.pdf.

109. Ibid.

110. P. R. Freeman and A. O'Hagan, "Thomas Bayes's Army [The Battle Hymn of Las Fuentes]," in *The Bayesian Songbook*, ed., Carlin and Bradley,

Yumpu, 2006, 37, https://www.yumpu.com/en/document/read/11717939/the-bayesian-songbook-university-of-minnesota.

111. Professor Sir David Spiegelhalter, Twitter, 2022, https://twitter.com/d_spiegel/status/1555822628996259840.

112. Ibid.

113. Aubrey Clayton, personal conversation, 2022.

114. Maurice Kendall and Alan Stuart, *The Advanced Theory of Statistics* (London: Charles Griffin, 1960).

115. M. G. Kendall, "On the Future of Statistics—A Second Look," *Journal of the Royal Statistical Society*, Series A (General) 131, no. 2 (1968): 182–204.

116. D. V. Lindley, "The Future of Statistics: A Bayesian 21st Century," *Advances in Applied Probability* 7 (1975): 106–15, https://doi.org/10.2307/1426315.

117. Larry Wasserman, "Is Bayesian Inference a Religion?," *Normal Deviate*, September 1, 2013, https://normaldeviate.wordpress.com/2013/09/01/is-bayesian-inference-a-religion/.

118. "Breathing Some Fresh Air Outside of the Bayesian Church," *Bayesian Kitchen*, December 5, 2013, http://bayesiancook.blogspot.com/2013/12/breathing-some-fresh-air-outside-of.html.

119. G. E. P. Box, "An Apology for Ecumenism in Statistics," in *Scientific Inference, Data Analysis, and Robustness*, ed., G. E. P. Box, T. Leonard, and C. F. J. Wu (Cambridge, MA: Academic Press,1983), 51–84.

CHAPTER TWO: BAYES IN SCIENCE

1. Diederik Stapel, *Onderzoek de psychologie van vlees*, Marcel Zeelenberg and Roos Vonk (2011).

2. D. A. Stapel and S. Lindenberg, "Coping with Chaos: How Disordered Contexts Promote Stereotyping and Discrimination," *Science* 332, no. 6026 (2011): 251–53.

3. Yudhijit Bhattacharjee, "The Mind of a Con Man," *New York Times*, April 28, 2013, https://www.nytimes.com/2013/04/28/magazine/diederik-stapels-audacious-academic-fraud.html.

4. D. J. Bem, "Feeling the Future: Experimental Evidence for Anomalous Retroactive Influences on Cognition and Affect," *Journal of Personality and Social Psychology* 100, no. 3 (March 2011): 407–25, doi: 10.1037/a0021524, PMID: 21280961.

5. J. A. Bargh, M. Chen, and L. Burrows, "Automaticity of Social Behavior: Direct Effects of Trait Construct and Stereotype-Activation on Action," *Journal of Personality and Social PsycholoIgy* 71, no. 2 (August 1996): 230–44, doi: 10.1037/0022-3514.71.2.230, PMID: 8765481.

6. K. D. Vohs, N. L. Mead, and M. R. Goode, "The Psychological Consequences of Money," *Science* 314, no. 5802 (November 17, 2006): 1154–56, doi: 10.1126/science.1132491, erratum in *Science* 349, no. 6246 (July 24, 2015): aac9679, PMID: 17110581.

7. S. W. Lee and N. Schwarz, "Bidirectionality, Mediation, and Moderation of Metaphorical Effects: The Embodiment of Social Suspicion and Fishy Smells," *Journal of Personality and Social Psycholology* 103, no. 5 (November 2012): 737–49, doi: 10.1037/a0029708, epub August 20, 2012, PMID: 22905770.

8. Daniel Kahneman, *Thinking, Fast and Slow* (New York: Farrar, Straus and Giroux 2011), 56–57.

9. J. P. Simmons, L. D. Nelson, and U. Simonsohn, "False-Positive Psychology: Undisclosed Flexibility in Data Collection and Analysis Allows Presenting Anything as Significant," *Psychological Science* 22, no. 11 (2011): 1359–66, doi:10.1177/0956797611417632.

10. J. P. Ioannidis, "Why Most Published Research Findings Are False," *PlOS Medicine* 2, no. 8 (August 2005): e124, doi: 10.1371/journal.pmed.0020124.

11. D. V. Lindley, "Sir Harold Jeff Reys," *Chance* 4, no. 2 (1991): 10–21, doi:10.1080/09332480.1991.11882423.

12. Malte Elson, "FlexibleMeasures.com: Competitive Reaction Time Task," Flexible Measures, 2016, http://www. flexiblemeasures.com/crtt/ https://doi.org/10.17605/OSF.IO/4G7FV.

13. K. M. Kniffin, O. Sigirci, and B. Wansink, "Eating Heavily: Men Eat More in the Company of Women," *Evolutionary Psychological Science* 2 (2016): 38–46, https://doi.org/10.1007/s40806-015-0035-3.

14. B. Wansink et al., "Attractive Names Sustain Increased Vegetable Intake

in Schools," *Preventive Medicine* 55, no. 4 (October 2012): 330–32, doi: 10.1016/j.ypmed.2012.07.012.

15. Brian Wansink, "The Grad Student Who Never Said 'No,'" 2016, archived at https://archive.ph/cPxmm.

16. Stephanie M. Lee, "Here's How Cornell Scientist Brian Wansink Turned Shoddy Data into Viral Studies about How We Eat," *BuzzFeed News*, February 26, 2018, https://www.buzzfeednews.com/article/stephaniemlee /brian-wansinkcornell-p-hacking.

17. Retraction Watch database: http://retractiondatabase.org/Retraction Search.aspx?AspxAutoDetectCookieSupport=1#?AspxAutoDetectCookie Support%3d1%26auth%3dWansink%252c%2bBrian.

18. Stephanie M. Lee, "Cornell Just Found Brian Wansink Guilty of Scientific Misconduct and He Has Resigned," *BuzzFeed News*, September 20, 2018, https://www.buzzfeednews.com/article/stephaniemlee/brian-wansink-re tired-cornell.

19. D. J. Bem, "Writing the Empirical Journal Article," in *The Compleat Academic: A Practical Guide for the Beginning Social Scientist*, ed., M. Zanna and J. Darley (New York: Random House, 1987), 171–201.

20. Open Science Collaboration, "Estimating the Reproducibility of Psychological Science," *Science* 349, no. 6251 (August 28, 2015): aac4716, doi: 10.1126/science.aac4716, PMID: 26315443.

21. H. Haller and S. Kraus, "Misinterpretations of Significance: A Problem Students Share with Their Teachers?," *Methods of Psychological Research* 7, no. 1 (2002): 1–20.

22. S. A. Cassidy et al., "Failing Grade: 89% of Introduction-to-Psychology Textbooks That Define or Explain Statistical Significance Do So Incorrectly," *Advances in Methods and Practices in Psychological Science* 2, no. 3 (2019): 233–39, https://doi.org/10.1177/2515245919858072.

23. Giulia Brunetti, "Neutrino Velocity Measurement with the OPERA Experiment in the CNGS Beam," *Journal of High Energy Physics* (October 2012): doi: 10.1007/JHEP10(2012)093.

24. Matt Strassler, "OPERA: What Went Wrong," Of Particular Significance, April 2, 2012, https://profmattstrassler.com/articles-and-posts/particle -physics-basics/neutrinos/neutrinos-faster-than-light/opera-what-went -wrong/.

25. David Hume, *An Enquiry Concerning Human Understanding*, Section IV,

Part II.28, reprinted from the Posthumous Edition of 1777, and edited with introduction, comparative tables of contents, and analytical index by L. A. Selby-Bigge, M.A., late fellow of University College, Oxford, 2nd ed. (1902).

26. Ibid., Part I.36.

27. Paul Feyerabend, "From Incompetent Professionalism to Professionalized Incompetence—the Rise of a New Breed of Intellectuals," *Philosophy of Social Science* 8 (1978): 37–53.

28. Karl Popper, *Realism and the Aim of Science* (London: Routledge, 1985), Chapter I, Section 3, I.

29. Karl Popper, *The Logic of Scientific Discovery* (London: Routledge, 2002 pbk, 2005 ebook), 91.

30. Karl Popper, *Realism and the Aim of Science: From the Postscript to the Logic of Scientific Discovery* (London: Routledge, 1985).

31. Michael Evans, *Measuring Statistical Evidence Using Relative Belief* (Boca Raton, FL: CRC Press, 2015), 107.

32. Johnny van Doorn et al., "Strong Public Claims May Not Reflect Researchers' Private Convictions," PsyArXiv, October 7, 2020.

33. Einstein, letter cited in C. Howson and P. Urbach, *Scientific Reasoning: The Bayesian Approach* (Chicago: Open Court, 1989), 7.

34. Einstein, quoted in Abraham Pais, *Subtle Is the Lord: The Science and the Life of Albert Einstein* (Oxford, UK: Oxford University Press, 1982), 159, cited in Howson and Urbach, *Scientific Reasoning*, 7.

35. Daniël Lakens, "Improving Your Statistical Inferences," Coursera, 3.2: Optional Stopping, https://www.coursera.org/learn/statistical-inferences /supplement/SES3h/assignment-3-2-optional-stopping.

36. D. V. Lindley, "A Statistical Paradox," *Biometrika* 44 (1957): 187–92.

37. W. Edwards, H. Lindman, and L. J. Savage, "Bayesian Statistical Inference for Psychological Research," *Psychological Review* 70 (1963): 193–242.

38. E. J. Wagenmakers et al., "An Agenda for Purely Confirmatory Research," *Perspectives on Psychological Science* 7 (2012): 627–33.

39. J. N. Rouder, "Optional Stopping: No Problem for Bayesians," *Psychonomic Bulleting and Review* 21 (2014): 301–8, https://doi.org/10.3758 /s13423-014-0595-4.

40. D. Bakan, "The Test of Significance in Psychological Research," *Psychological Bulletin* 66, no. 6 (1966): 423–37, doi: 10.1037/h0020412.

41. P. E. Meehl, "Why Summaries of Research on Psychological Theories Are Often Uninterpretable," *Psychological Reports* 66, no. 1 (1990): 195–244, https://doi.org/10.2466/PR0.66.1.195-244.

42. Lindley, "A Statistical Paradox."

43. D. J. Benjamin et al., "Redefine Statistical Significance," *Nature Human Behaviour* 2 (2018): 6–10, https://doi.org/10.1038/s41562-017-0189-z.

44. Cassie Kozyrkov, "Statistics: Are You Bayesian or Frequentist?," Towards Data Science, June 4, 2021, https://towardsdatascience.com/statistics -are-you-bayesian-or-frequentist-4943f953f21b.

45. Suetonius, *De vita Caesarum*, lib. I, xxxii.

46. Population on January 1, 2022, Eurostat Data Browser, https://ec.europa .eu/eurostat/databrowser/view/tps00001/default/table?lang=en.

47. Cited by Andrew Gelman, "If You're Not Using a Proper, Informative Prior, You're Leaving Money on the Table," *Statistical Modeling, Causal Inference, and Social Science*, November 21, 2014, https://statmodeling.stat .columbia.edu/2014/11/21/youre-using-proper-informative-prior-youre -leaving-money-table/.

48. Kozyrkov, "Statistics."

CHAPTER THREE: BAYESIAN DECISION THEORY

1. Aristotle (4th century BC), *Physics*, translation with commentary by H. G. Apostle (Bloomington: Indiana University Press, 1969).

2. E. T. Jaynes, *Probability Theory: The Logic of Science*, ed., G. Bretthorst (Cambridge, UK: Cambridge University Press, 2003), doi: 10.1017/CB O9780511790423.

3. G. Boole, *An Investigation of the Laws of Thought, on Which Are Founded the Mathematical Theories of Logic and Probabilities* (London: Walton and Maberly, 1854), reprinted as *George Boole's Collected Works*, vol. 2 (1916; New York: Dover, 1951).

4. Jaynes, *Probability Theory*, 3.

5. Eliezer Yudkowsky, *Rationality: From AI to Zombies* (Berkeley, CA: Machine Intelligence Research Institute, 2015), 104.

6. Ibid., 104.

7. Ibid., 792, 202.

8. Jaynes, *Probability Theory*, 35.

9. Oliver Cromwell, Letter 129, 1650, http://www.olivercromwell.org/Letters_and_speeches/letters/Letter_129.pdf.

10. Dennis Lindley, *Making Decisions*, 2nd ed. (New York: Wiley, 1991), 104.

11. Yudkowsky, *Rationality*, 245.

12. "How NICE Measures Value for Money in Relation to Public Health Interventions," NICE, September 1, 2013, https://www.nice.org.uk/media/default/guidance/lgb10-briefing-20150126.pdf.

13. "List of Things Named after John von Neumann," Wikipedia, https://en.wikipedia.org/wiki/List_of_things_named_after_John_von_Neumann.

14. Much of the following is drawn from Ananyo Bhattacharya, *The Man from the Future: the Visionary Life of John von Neumann* (New York: W. W. Norton, 2022), 160.

15. J. von Neumann and O. Morgenstern, *Theory of Games and Economic Behavior*, 6th printing (Princeton, NJ: Princeton University Press, 1955), 10.

16. Eliezer Yudkowsky, "Occam's Razor," Read the Sequences, 2015, 115, https://www.readthesequences.com/Occams-Razor.

17. Michal Koucký, "A Brief Introduction to Kolmogorov Complexity, University Karlova, May 4, 2006", http://iuuk.mff.cuni.cz/~koucky/vyuka/ZS2013/kolmcomp.pdf.

18. This example is taken from Jaynes, *Probability Theory*, 4.

19. S. G. Soal, "Fresh Light on Card Guessing: Some New Effects," *Proceedings of the Society for Psychical Research* 46 (1940): 152–98.

20. Stuart Russell and Peter Norvig, *Artificial Intelligence: A Modern Approach*, 3rd ed. (Chennai, India: Pearson, 2010), 9.

21. Most of this section is taken from a conversation I had with Dr. William Woof of UCL, who uses AI and machine learning techniques to improve diagnosis of retinal diseases.

22. Emily M. Bender et al., "On the Dangers of Stochastic Parrots: Can Language Models Be Too Big?," in *FAccT '21: Proceedings of the 2021 ACM Conference on Fairness, Accountability, and Transparency* (New York: Association for Computing Machinery, 2021), 610–23, https://doi.org/10.1145/3442188.3445922.

23. Kenneth Li et al., "Emergent World Representations: Exploring a Se-

quence Model Trained on a Synthetic Task," ArXiv, Cornell University, last updated February 27, 2023, https://doi.org/10.48550/arXiv.2210 .13382.

24. Kenneth Li, "Do Large Language Models Learn World Models or Just Surface Statistics?," Gradient, January 21, 2023.

25. Belinda Z. Li, Maxwell Nye, and Jacob Andreas, "Implicit Representations of Meaning in Neural Language Models," ArXiv, Cornell University, 2021, https://doi.org/10.48550/arXiv.2106.00737.

CHAPTER FOUR: BAYES IN THE WORLD

1. S. Lichtenstein et al., "Judged Frequency of Lethal Events," *Journal of Experimental Psychology: Human Learning and Memory* 4, no. 6 (1978): 551–78, https://doi.org/10.1037/0278-7393.4.6.551.

2. Amos Tversky and Daniel Kahneman, "Judgments of and by Representativeness," in *Judgment Under Uncertainty: Heuristics and Biases*, ed., Daniel Kahneman, Paul Slovic, and Amos Tversky (New York: Cambridge University Press, 1982), 96.

3. Amos Tversky and Daniel Kahneman, "The Framing of Decisions and the Psychology of Choice," *Science* 211, no. 4481 (January 30, 1981): 453–58, doi: 10.1126/science.7455683. PMID: 7455683.

4. Retraction for Shu et al., "Signing at the Beginning Makes Ethics Salient and Decreases Dishonest Self-Reports in Comparison to Signing at the End," *Proceedings of the National Academy of Sciences USA* 109, no. 38 (August 27, 2021):15197–200, doi: 10.1073/pnas.1209746109.

5. Cathleen O'Grady, "Fraudulent Data Raise Questions about Superstar Honesty Researcher," *Science* (August 24, 2021), https://www.science.org /content/article/fraudulent-data-set-raise-questions-about-superstar-hon esty-researcher.

6. W. Casscells, A. Schoenberger, and T. B. Graboys, "Interpretation by Physicians of Clinical Laboratory Results," *New England Journal of Medicine* 299, no. 18 (1978): 999–1001.

7. B. L. Anderson, S. Williams, and J. Schulkin, "Statistical Literacy of Obstetrics-Gynecology Residents," *Journal of Graduate Medical Education* 5, no. 2 (June 2013): 272–75, doi: 10.4300/JGME-D-12-00161.1.

8. P. C. Wason, "Reasoning about a Rule," *Quarterly Journal of Experimental Psychology* 20, no. 3 (1968): 273–81, doi: 10.1080/14640746808400161.

9. Jonathan St. B. T. Evans et al., *Human Reasoning: The Psychology of Deduction* (London: Psychology Press, 1993).

10. L. Cosmides and J. Tooby, "Cognitive Adaptions for Social Exchange," in *The Adapted Mind: Evolutionary Psychology and the Generation of Culture*, ed. J. Barkow, L. Cosmides, and J. Tooby (New York: Oxford University Press, 1992), 163–228.

11. Louis Liebenberg, personal communication, cited in Steven Pinker, *Rationality: What It Is, Why It Seems Scarce, Why It Matters* (New York: Viking, 2021), 4.

12. J. K. Madsen, "Trump Supported It?! A Bayesian Source Credibility Model Applied to Appeals to Specific American Presidential Candidates' Opinions," in A. Papafragou et al., eds., *Proceedings of the 38th Annual Conference of the Cognitive Science Society*, 2016, 165–70.

13. Douglas Adams, *Dirk Gently's Holistic Detective Agency* (New York: Simon & Schuster, 1987), 153.

14. Gerd Gigerenzer and Henry Brighton, "Homo Heuristicus: Why Biased Minds Make Better Inferences," *Topics in Cognitive Science* 1, no. 1, (2009): 107–43, doi: 10.1111/j.1756-8765.2008.01006.x, hdl: 11858/00-001M-0000-0024-F678-0.

15. Dennis Shaffer et al., "How Dogs Navigate to Catch Frisbees," *Psychological Science* 15 (2004): 437–41, doi: 10.1111/j.0956-7976.2004.00698.x.

16. R. P. Hamlin, " 'The Gaze Heuristic': Biography of an Adaptively Rational Decision Process," *Topics in Cognitive Science* 9 (2017): 264–88, doi: 10.1111/tops.12253.

17. Garrick Blalock, Vrinda Kadiyali, and Daniel Simon, "Driving Fatalities After 9/11: A Hidden Cost of Terrorism," *Applied Economics* 41 (2009): 1717–29, doi: 10.1080/00036840601069757.

18. "Letters to the Editor," *American Statistician* 29, no. 1 (1975): 67–71, doi: 10.1080/00031305.1975.10479121.

19. Marilyn vos Savant, "Game Show Problem," *Parade*, 1990.

20. Andrew Vazsonyi, *Which Door Has the Cadillac: Adventures of a Real-Life Mathematician*, (Lincoln, NE: iUniverse, 2002), 5.

21. Martin Gardner, *The Second Scientific American Book of Mathematical Puzzles and Diversions* (New York: Simon & Schuster, 1961).

22. John Lewis Gaddis, *The Cold War: A New History* (New York: Penguin Press, 2005), 228.

23. Ibid.

24. Philip Tetlock and Dan Gardner, *Superforecasting* (New York: Crown, 2015), 50.

25. Andrew Mauboussin and Michael J. Mauboussin, "If You Say Something Is 'Likely,' How Likely Do People Think It Is?," *Harvard Business Review*, July 3, 2018, https://hbr.org/2018/07/if-you-say-something-is-likely-how -likely-do-people-think-it-is.

26. Tetlock and Gardner, *Superforecasting*, 59.

27. Ibid., 73.

28. Ibid., 113.

29. Ibid., 157.

30. *Jacobellis*, 378 U.S. at 197 (Stewart, J., concurring).

31. Ludwig Wittgenstein, *Philosophical Investigations* (Hoboken, NJ: Wiley-Blackwell, 1953), 7.

32. Diogenes Laërtius, "The Cynics: Diogenes," *Lives of the Eminent Philosophers*, vol. 2, trans. Robert Drew Hicks (London: Loeb Classical Library, 1925), 6.

CHAPTER FIVE: THE BAYESIAN BRAIN

1. Plato, *The Republic*, Book VII, ed., W. H. D. Rouse (New York: Penguin Group 1951), 365–401.

2. Sylvia Berryman, "Democritus," in *The Stanford Encyclopedia of Philosophy*, ed., Edward N. Zalta, Winter 2016 ed., https://plato.stanford.edu /archives/win2016/entries/democritus/.

3. Vasco Ronchi, *Nature of Light: An Historical Survey* (London: Heinemann, 1970), 16.

4. Ibn al-Haytham, *Book of Optics*, book 1, ch. 32, trans. A. I. Sabra (London: Warburg Institute, 1989), https://monoskop.org/images/f/ff/The_Optics _of_Ibn_Al-Haytham_Books_I-III_On_Direct_Vision_Sabra_1989.pdf.

5. Immanuel Kant, W. S. Pluhar, and P. Kitcher, *Critique of Pure Reason* (1787; Indianapolis, IN: Hackett, 1996).

6. L. R. Swanson, "The Predictive Processing Paradigm Has Roots in Kant," *Frontiers in Systems Neuroscience* 10, no. 79 (October 10, 2016), doi: 10.3389/fnsys.2016.00079. PMID: 27777555; PMCID: PMC5056171.

7. Hermann von Helmholtz, "*Vorläufiger Bericht über die Fortpflanzungs-Geschwindigkeit der Nervenreizung*," *Archiv für Anatomie, Physiologie und wissenschaftliche Medicin* (1850): 71–73.

8. Hermann von Helmholtz, "The Recent Progress of the Theory of Vision," 1868, in *Science and Culture: Popular and Philosophical Essays*, ed., David Canahan (Chicago: University of Chicago Press, 1995), 127–203.

9. R. L. Gregory, *Eye and Brain*, 5th ed. (Oxford, UK: Oxford University Press, 1998).

10. Ibid.

11. Cates Holderness, "What Colors Are This Dress?," *BuzzFeed*, February 26, 2015, https://www.buzzfeed.com/catesish/help-am-i-going-insane-its -definitely-blue.

12. *Philosophical Transactions of the Royal Society of London, Series B* 352 (1997): 1121–28.

13. S. Aston and A. Hurlbert, "What #theDress Reveals about the Role of Illumination Priors in Color Perception and Color Constancy," *Journal of Visualized Experiments* 17, no. 9 (August 1, 2017): 4, doi: 10.1167/17.9.4, PMID: 28793353, PMCID: PMC5812438.

14. R. Fitzhugh, "A Statistical Analyzer for Optic Nerve Messages," *Journal of General Physiology*, 41, no. 4 (March 20, 1958): 675–92, doi: 10.1085 /jgp.41.4.675, PMID: 13514004, PMCID: PMC2194875.

15. This account is largely taken from Andy Clark's *Surfing Uncertainty: Prediction, Action, and the Embodied Mind* (Oxford, UK: Oxford University Press, 2015) and from Anil Seth, *Being You: A New Science of Consciousness* (London: Faber & Faber, 2021).

16. Marc Ernst and Martin Banks, "Humans Integrate Visual and Haptic Information in a Statistically Optimal Fashion," *Nature* 415 (2002): 429–33, doi: 10.1038/415429a.

17. See "McGurk Effect—Auditory Illusion," *Horizon*, BBC video, https:// www.youtube.com/watch?v=2k8fHR9jKVM.

18. "Green Needle or Brainstorm?," Illinois Vision Lab, https://publish.illinois .edu/visionlab/2021/01/27/green-needle-brainstorm/.

19. W. Schultz, "Reward Signaling by Dopamine Neurons," *Neuroscientist* 7, no. 4 (August 2001): 293–302, doi: 10.1177/107385840100700406, PMID: 11488395.

20. For instance: R. P. N. Rao and T. J. Sejnowski, "Predictive Coding, Cortical Feedback, and Spike-Timing Dependent Plasticity," in *Probabilistic Models of the Brain: Perception and Neural Function*, ed., R. P. N. Rao, B. A. Olshausen, and M. S. Lewicki (Cambridge, MA: MIT Press, 2002), 297–315.

21. T. Hosoya, S. Baccus, and M. Meister, "Dynamic Predictive Coding by the Retina," *Nature* 436 (2005): 71–77, https://doi.org/10.1038/nature03689.

22. A. Kolossa, B. Kopp, and T. Fingscheidt, "A Computational Analysis of the Neural Bases of Bayesian Inference," *Neuroimage* 106 (February 1, 2015): 222–37, doi: 10.1016/j.neuroimage.2014.11.007, epub November 8, 2014, PMID: 25462794.

23. Benjamin W. Tatler et al., "Eye Guidance in Natural Vision: Reinterpreting Salience," *Journal of Vision* 11, no. 5 (2011): 5, doi: https://doi.org/10.1167/11.5.5, cited in Clark, *Surfing Uncertainty*, 67.

24. "The Saccadic Tracking Loop," Fault Tolerant Tennis, September 20, 2022, https://faulttoleranttennis.com/the-saccadic-tracking-loop/.

25. M. F. Land and B. W. Tatler, *Looking and Acting: Vision and Eye Movements in Natural Behaviour* (Oxford, UK: Oxford University Press, 2009), https://doi.org/10.1093/acprof:oso/9780198570943.001.0001.

26. Sarah-Jayne Blakemore, Daniel Wolpert, and Chris Frith, "Why Can't You Tickle Yourself?," *Neuroreport* 11, no. 11 (August 2000): 11–16, doi: 10.1097/00001756-200008030-00002.

27. S. S. Shergill et al., "Two Eyes for an Eye: The Neuroscience of Force Escalation," *Science* 301, no. 5630 (July 11, 2003): 187, doi: 10.1126/science.1085327, PMID: 12855800.

28. S. J. Blakemore, D. M. Wolpert, and C. D. Frith, "Central Cancellation of Self-Produced Tickle Sensation," *Nature Neuroscience* 1, no. 7 (November 1998): 635–40, doi: 10.1038/2870, PMID: 10196573.

29. Chris Frith, *Making Up the Mind: How the Brain Creates Our Mental World* (Hoboken, NJ: Blackwell, 2007), 102.

30. H. M. Wichowicz et al., "Hollow Mask Illusion—Is It Really a Test for Schizophrenia?," *Psychiatria Polska* 50, no. 4 (2016): 741–45, doi: 10.12740/PP/60150, PMID: 27847925.

31. Frith, *Making Up the Mind*, 108.

32. R. Carhart-Harris et al., "Trial of Psilocybin versus Escitalopram for Depression," *New England Journal of Medicine* 384, no. 15 (April 15, 2021): 1402–11, doi: 10.1056/NEJMoa2032994, PMID: 33852780.

33. A. K. Davis et al., "Effects of Psilocybin-Assisted Therapy on Major Depressive Disorder: A Randomized Clinical Trial," *JAMA Psychiatry* 78, no. 5 (2021): 481–89, doi: 10.1001/jamapsychiatry.2020.3285; C. S. Grob et al., "Pilot Study of Psilocybin Treatment for Anxiety in Patients with Advanced-Stage Cancer," *Archives of General Psychiatry* 68, no. 1 (2011): 71–78, doi: 10.1001/archgenpsychiatry.2010.116; S. Ross et al., "Rapid and Sustained Symptom Reduction Following Psilocybin Treatment for Anxiety and Depression in Patients with Life-Threatening Cancer: A Randomized Controlled Trial," *Journal of Psychopharmacology* 30, no. 12 (2016): 1165–80, doi: 10.1177/0269881116675512; R. R. Griffiths et al., "Psilocybin Produces Substantial and Sustained Decreases in Depression and Anxiety in Patients with Life-Threatening Cancer: A Randomized Double-Blind Trial," *Journal of Psychopharmacology* 30, no. 12 (2016): 1181–97, doi: 10.1177/0269881116675513.

34. Scott Alexander, "God Help Us, Let's Try to Understand Friston on Free Energy," *Slate Star Codex*, March 4, 2018, https://slatestarcodex.com/2018/03/04/god-help-us-lets-try-to-understand-friston-on-free-energy/.

CONCLUSION: BAYESIAN LIFE

1. J. Clark, S. Watson, and K. Friston, "What Is Mood? A Computational Perspective," *Psychological Medicine* 48, no. 14 (2018): 2277–84, doi: 10.1017/S0033291718000430; P. R. Corlett, C. D. Frith, and P. C. Fletcher, "From Drugs to Deprivation: A Bayesian Framework for Understanding Models of Psychosis," *Psychopharmacology* 206, no. 4 (November 2009): 515–30, doi: 10.1007/s00213-009-1561-0, epub May 28, 2009, PMID: 19475401; PMCID: PMC2755113.

2. Fred Hoyle, *The Intelligent Universe: A New View of Creation and Evolution*, (London: Michael Joseph, 1983), 19.

3. J. Huber et al., "Nobel and Novice: Author Prominence Affects Peer Review," *Proceedings of the National Academy of Science USA* 119, no. 41 (Oc-

tober 11, 2022): e2205779119, doi: 10.1073/pnas.2205779119, epub October 4, 2022, PMID: 36194633, PMCID: PMC9564227.

4. George E. P. Box, "Science and Statistics," *Journal of the American Statistical Association* 71, no. 356 (1976): 791–99, doi: 10.1080 /01621459.1976.10480949.

Index

Page numbers followed by * refer to footnotes.

About the Author

Tom Chivers is an author and the science writer for *Semafor*. He won the Association of British Science Writers' award for Science Journalist of the Year in 2021, and the Royal Statistical Society's Statistical Excellence in Journalism Award in 2018 and in 2020. He's written for the London *Times*, the *i*, the *Telegraph*, the *Observer*, the *Guardian*, politics.co.uk, *New Scientist*, CNN, *Wired*, *Smithsonian Air & Space*, and others. His previous books include *The Rationalist's Guide to the Galaxy* and *How to Read Numbers*.